HOW BIRDS BEHAVE

Interpreting what they do and why

A catalogue record for this book is available from the National Library of Australia.

ISBN: 9781486313280 (hbk)

This edition published exclusively in Australia and New Zealand by:

CSIRO Publishing, Locked Bag 10
Clayton South VIC 3169, Australia
Telephone: +61 3 9545 8400
publishing.sales@csiro.au
www.publish.csiro.au

For my parents

Publisher: Nigel Browning
Commissioning editor: Kate Shanahan
Project manager: Kate Duffy
Design and art direction: Wayne Blades
Picture researcher: Katie Greenwood
Illustrators: Kate Osborne and John Woodcock

Front cover image: Juan Carlos Munoz/Nature Picture Library

Printed in China

10 9 8 7 6 5 4 3 2 1

Great egrets fighting in midair.

HOW BIRDS BEHAVE

Interpreting what they do and why

WENFEI TONG

Foreword by **Ben Sheldon**

PUBLISHING

A flock of yellow-headed blackbirds.

CONTENTS

FOREWORD

BIRDS SEEM, AT FIRST SIGHT, a living paradox of adaptability. Compared to any other class of animals, they are remarkably invariant. We don't need much information to learn what a bird is: they all have two legs, two wings (even if these are sometimes tiny and non-functional), a beak with which they manage all of their dietary requirements and feathers. Compare this similarity in body plan with the diversity shown by mammals, or any other class of vertebrates. In other aspects of their morphology, birds are also remarkably invariant. The range in size from the smallest (bee hummingbird) to the largest (ostrich) species of bird seems impressive, but it is dwarfed by the range in mammals from shrew to blue whale, or in fish from dwarf pygmy goby to whale shark. Equally, birds seem boringly similar in that they all lay eggs, compared to the diversity of reproductive systems in reptiles, fishes, amphibians and even mammals.

Despite this limited variation, birds exploit a wider range of the biosphere than any other animal class, with records of birds from the North and South Poles, ranging across all of the world's oceans, breeding on the remotest islands, and found in the hottest and driest deserts. Alpine choughs have been recorded breeding at altitudes of over 6,000 metres (20,000 feet) in the Himalayas, while house sparrows, which live in close association with human settlements worldwide, have been recorded in the UK breeding 640 metres (2,100 feet) underground in a Yorkshire coal mine, subsisting on crumbs provided by miners. Emperor penguins have been logged diving to 550 metres (1,800 feet) in Antarctic oceans, and Ruppell's vultures reported at altitudes of over 11 kilometres (km) (6.8 miles), higher than many aeroplanes.

The key to this remarkable success and global distribution, despite all birds being based on very much the same body plan, is often said to be the remarkable flexibility that the evolution of feathers afforded early birds. While this is one perspective, morphological adaptations are only part of the story, and a limited part, at that, without behavioural adaptations to exploit them. Birds share many features with us that have made them wonderful subjects for understanding evolution, diversity and behavioural adaptation. Many species are conspicuous, and most active during daylight hours; vision and hearing are key sensory modes that birds exploit in communication, just as we do. Most birds even have reproductive life histories rather like ours: a pair construct a nest (home), and rear a family over an extended period. In some cases, there are even 'divorces' and the avian equivalent of extra-marital affairs.

This book is a dazzling tapestry woven of stories of the many ways in which bird behaviour enables us to understand adaptation. Wenfei Tong's lively prose, scattered with personal anecdotes and experiences, weaves a compelling picture, mixing classic examples with the very latest research findings, which continue to expand our knowledge of the diversity and richness of bird behaviour. Remarkable adaptations described here come from exotic and extreme environments, but also from birds that will be familiar to the urban birdwatcher worldwide. I hope that this diversity and accessibility will inspire a new generation of research in this area.

BEN SHELDON

INTRODUCTION

A CASCADE OF LIQUID NOTES made me look up from the grey New York street to admire an iridescent, green and purple bird with tiny white arrowheads highlighting the tips of each glistening feather. The European starling is so widespread it can easily go unappreciated, but this bird exemplifies much of what this book is about – from why birds behave the way they do, to how their interactions with each other and with humans inspire and influence our view of life.

European starlings appear repeatedly in this book, partly because they feature so much in both art and science. Shakespeare's only known reference to the species, in *Henry IV, Part I*, is to the starling's ability to mimic almost anything. There are scholarly speculations, including a long essay by biologists who specialise in the study of birdsong, about how Mozart and his pet starling inspired each other musically. We do know that Mozart loved his little 'joker', going as far as to write an elegy to the bird when it died. European starlings are so tractable, adaptable and social that they are a beloved species for biologists studying everything from language, to economic decision-making and to collective animal motion.

NATURE VIA NURTURE

When you watch a bird constructing an intricate nest or screaming in alarm at a hawk, you might well wonder how it 'knows' what to do and when. In fact, most bird behaviour is a combination of innate neural hardwiring, and constant adjustments and reprogramming in response to its environment. For instance, a young bird may be predisposed, through generations of natural selection, to emit a certain sound when alarmed. However, it must learn how to refine the alarm call and when to sound it by observing adults attacking a predator. A European starling has the capacity to produce a perfect rendition of a red-tailed hawk's scream, or 'I love you' in multiple human languages, but only if it is exposed to those sounds.

Above

I watched these lesser masked weavers constructing their nests every day on a friend's farm in Zambia. This male is putting the finishing touches to its nest.

Above

There are few sounds that so evocatively epitomise a particular landscape as the song of the western meadowlark, which speaks to me of Big Sky Country in the American West.

AESTHETICS, ANTHROPOMORPHISM AND AMORALITY

To what extent can humans and birds really identify with and understand each other? This depends rather a lot on the individuals in question, but birds have a lot to teach us in terms of how other species have evolved to perceive the world. European starlings often make more rational (as defined by economic theory) decisions than humans when foraging, as discussed in chapter 1.

As a gregarious species, they serve as a popular model for how songbirds and humans learn to communicate vocally – see chapter 2. In chapter 3, on courtship, we learn how the starling got its beautiful feathers and virtuosic vocabulary by selecting the most attractive mates. Chapter 4 deals with raising a family. Female starlings exposed to more predation can 'programme' their chicks to be better at eluding predators because eggs exposed to higher levels of stress hormone hatch into chicks with better-developed flight muscles. This also illustrates one of many ways birds deal with danger in chapter 5, where we also explore how and why starlings form such massive and coordinated flocks, called murmurations. In the last chapter, on climate, we see how highly flexible birds, such as European starlings, cope easily with a changing environment, expanding their range and often ceasing to migrate when settling down in a city all year round is the sensible thing to do.

The European starling came to the New World in the 1890s, when a well-intentioned member of the American Acclimatization Society decided to introduce all the birds mentioned in Shakespeare's plays, and released about sixty starlings into New York's Central Park. These highly adaptable birds now constitute a major introduced pest in North America, wending their way north and west as far as Alaska, especially as human development and climate change continue to make the frozen north ever more salubrious.

The starling's success is the perfect example of how evolution is amoral. It would be unfair to blame these consummate opportunists for the fact that they cost the US about $1 billion a year in crop damage, or that their success as immigrants may be threatening endangered natives such as the red-headed woodpecker.

If some of the language in this book smacks of anthropomorphism, that is because as a social species, it can be tricky, even for scientists, to remain completely objective. Some words, such as 'divorce' or 'personality', are widespread in the technical literature, whereas others, such as 'extra-pair copulations' are more of a mouthful, and I will sometimes cut to the chase and use the human equivalent (extra-marital affairs).

BIRDS AS INDIVIDUALS

Biologists are starting to appreciate how much individual birds differ within a species, as well as how flexible individuals can be over the course of a lifetime. Testosterone influences how males switch between having more matings, and being better fathers to fewer children. Junco males with higher testosterone levels attract more females and have more extra-pair offspring. More testosterone doesn't increase the number of affairs male house sparrows or blue tits have, but they do attract more social mates. These males provide less paternal care than males with lower testosterone levels.

One of the things I love about birds, that most birdwatchers have in common, is the fact that they are everywhere, and provide an extra channel by which to experience a new place. This book celebrates the sheer diversity of bird behaviour from all over the world. By highlighting both classic and the most current scientific studies, it explores not just why birds behave the way they do, but how we know that. Ultimately, I hope to deepen your appreciation of the birds you encounter, wherever you are.

Below
I saw this purple-crowned fairy (hummingbird) having its breakfast while I had mine, in a beautiful garden in Costa Rica.

FINDING FOOD

Right

The anianiau is the smallest species among an evolutionary radiation of Hawaiian honeycreepers, each adapted to exploit a different food.

FEEDING FROM HEAD TO TAIL

Bird bills are tools. Every time you see a bird, you can guess what it eats just from the shape of its bill, although the rest of its form, where it is and how it behaves are additional clues. For example, seed specialists such as finches have much thicker bills than warblers, which hunt insects and spiders.

Below

A snail kite eats an apple snail in Panama.

Paying attention to bill size and shape is one of the ways experienced birders can proclaim so confidently that some small brown bird is a sparrow, not a wren. However, using a tool such as a bill to classify birds does have its pitfalls. While a young man, and before he came up with the theory of natural selection, the naturalist Charles Darwin (1809–82) assumed that the small birds he collected in the Galápagos came from at least four separate subfamilies, ranging from grosbeaks to warblers. He only learned that they were all finches from John Gould, an ornithologist at the British Museum, who examined and classified the specimens.

In the case of two famous island radiations – the Hawaiian honeycreepers (true finches) and Darwin's finches – skulls and bills change shape in concert much more than in other bird groups. This enhanced capacity to evolve bills rapidly to fit a particular function could explain why these groups diversified into so many species on Hawaii and the Galápagos, compared to other bird lineages that colonised the islands at the same time. However, not all tools evolve with equal speed. A study comparing the three-dimensional skull and bill shape of 352 species from across the evolutionary tree of birds found that some heads evolve faster than others, depending on diet. Seed and nectar feeders are the quickest to evolve different-shaped headgear. In contrast, raptors, such as eagles, hawks and owls, changed the most slowly. This could be because these birds of prey specialise on using their feet as weapons and on hunting strategy; their binocular vision and a good ripping bill can afford to remain largely unchanged.

Bucking this trend is a highly specialised raptor with an unusual diet. The bills of snail kites of Florida have adapted to extracting the meat from snail shells. In the mid-2000s, an exotic apple snail, two to five times larger than the kite's natural snail prey, invaded Florida. Counterintuitively, the kites, already highly endangered by habitat loss, have increased in numbers. They have benefited from the introduced food supply by adapting to the larger prey in a matter of two generations. This unusually fast change in bill size is partly due to strong natural selection for genes that make larger bills, and partly due to plastic bill development, where young birds fed on larger snails grow larger bills.

Tails can also dramatically change the physics of bird movement, which can be important for how they catch prey. Aquatic birds, in particular, have to optimise between flying and swimming to find food. Four unrelated groups of birds that feed underwater have converged on a terminal tailbone (a pygostyle) that is long and straight, to act as a rudder. These pygostyles differ further in shape, depending on whether the birds are plunge divers, such as gannets, paddlers, such as puffins, or wing-propelled swimmers, such as penguins. But, birds that fly or run for their food have shorter pygostyles.

DIVERSITY ENABLES COEXISTENCE

One way to explain the evolution and coexistence of diversity, be it in a human economy or in nature, is for some individuals to become specialists in a narrow 'niche'. Different foods can select for differently shaped bills and bodies among various bird species. In other cases, it is software, such as birds' behaviour, that changes more than the hardware.

Opposite

As with many popular bird groups, the common name 'warbler' is misleading. Old World warblers, such as the yellow-browed warbler (top), are part of an older evolutionary radiation, while the much more colourful wood warblers, such as the yellow warbler (bottom), evolved independently to fill many of the same ecological niches in the New World. In spite of these differences in ancestry, both warbler groups are examples of evolutionary radiations in which competition for food drove diversification, and so allowed multiple, closely related species to coexist.

Many birders will be familiar with the thrill of listening to a chorus of warblers, and trying to distinguish between the similarly shaped little birds all feeding in the same tree. A key question for biologists is how so many superficially similar species manage to thrive in close proximity.

Wood warblers of the Americas are an elegant example of the answer: niche diversification. Each warbler species has its own unique way of making a living, in its own part of the forest, or even in specialised parts of a tree. In the 1950s, the ecologist Robert H. MacArthur (1930–72) demonstrated this idea by painstakingly watching the feeding antics of five warbler species in the Maine woods. He spent hours accounting for exactly where, when and how each warbler was feeding among the spruce trees. A clear pattern of specialisation emerged, not in the warbler bills (which all look similar), but in their feeding behaviour.

If you divide a conveniently geometrical Christmas tree into sections, and pay close attention to the warblers, you should find that each species frequents a different section, and has a feeding style particularly suited to its unique place, in what Darwin called 'the economy of nature'.

WARBLER ECONOMICS

In MacArthur's study, the very tops of the spruce trees were dominated by Cape May and Blackburnian warblers. However, the Cape May warblers spent most of their time hawking for insects, flying away from the tree, and hanging about the outskirts of the treetops. In contrast, Blackburnian warblers scrutinised each branch for food, and generally stayed closer to the tree trunk. Even more conservative were the bay-breasted warblers. They skulked almost exclusively near the middle of each spruce, taking a long time to work their way systematically through the branches before moving on to the next tree. This species made the fewest feeding flights away from the trees.

Black-throated green warblers had a unique way of peering into the thick mat of spruce needles above them, before springing up and hovering to pick off their prey with pinpoint accuracy. This made for the most active, flighty feeding habits. MacArthur also noted that while the other warbler species fed silently, the black-throated green was a noisy eater, and uttered chirrups almost incessantly. The most general in habits and space-use were the yellow-rumped (Myrtle) warblers, which spent more time near the bottom of the trees than any of the other species, but could also be seen at the top. They hawked for insects like Cape May warblers, skulked like the Blackburnians and bay-breasteds, and moved the most from tree to tree.

Different warblers fill various foraging niches on a spruce tree

Cape May (**1**) and Blackburnian warblers (**2**) tend to forage near the treetop, while bay-breasted warblers (**3**) feed lower down, but closest to the tree trunk. Black-throated green warblers (**4**) have the most active, flighty feeding habits at mid-height, and while yellow-rumped (Myrtle) warblers (**5**) are often spotted near the bottom of a tree, they are also the most versatile in their feeding habits.

Birdwatchers repeating MacArthur's study today, even if they were in precisely the same spot, would likely not find exactly the same combinations of species occupying the same parts of a tree. This is because birds adapt, both as populations and as individuals. We know that there was a glut of spruce budworms in the 1950s, when MacArthur did his seminal PhD study, and that led to a population boom in the typically rare Cape May warblers.

The point is that niches are not static, they are flexible areas of expertise that can adapt to a dynamic feeding economy in which resources, competitors and the overall climate change constantly.

FROM SEXUAL SEGREGATION TO INDIVIDUAL SPECIALISATION

Some birds take niche differentiation further, with variation and specialisation minimising competition between members of the same species. A division of foraging labour between the sexes can be associated with differences in physique. For example, some species of male cormorants are larger than females, and can dive deeper and for longer. Giant petrel males bully females and exclude them from carrion, forcing the females to forage farther out at sea.

In contrast, there is no evidence of male domination among black-browed and grey-headed albatross. Both species do show marked spatial segregation between the sexes in their fishing grounds, but only during times of incubation, when foraging trips can be longer. Males of both species are larger than females, and need higher wind speeds to remain aloft. They are restricted to fishing in the windiest places, compared to their smaller mates.

Lest you think that sexually segregated feeding is restricted to birds that fish, green woodhoopoe males are also larger than females, and the sexes forage differently. Males have bills that are 36 per cent longer than those of females and specialise on probing under bark, whereas females spend most of their time pecking. This is likely to reduce competition between the sexes, because opposite sex pairs are more likely to feed together, and to bicker less. These feeding differences only emerge as the birds mature, as younger birds show no sex differences in bill shape.

Above

Green woodhoopoes are a garrulous and gregarious group of breeders from sub-Saharan Africa. Males develop longer bills than females, which helps to minimise competition for food between the sexes.

Among birds of prey the tables are turned, with males tending to be the smaller sex. In particular, the largest size differences between the sexes occur in raptors, where males benefit from being smaller and more agile. So, raptors with smaller prey that hunt in denser habitats, or that are more territorial than group-living, tend to have disproportionately small males. This could be because male raptors usually bear the brunt of provisioning during the breeding season, when females are sitting on eggs and chicks, and also because males often engage in aerial fights over territories.

While there is sex-specific specialisation among Eurasian oystercatchers, members of both sexes fall on a continuum. There is also substantial overlap, with many female and male oystercatchers having intermediate bills, which are decent at obtaining both types of prey. The birds on either end of the continuum have more of an edge in the coldest years, whereas in most years, birds of either sex with intermediate bills are likely to survive just as well as the extreme diet specialists. It is also unclear how much of this specialisation is driven by competition for resources as opposed to other factors, since females are usually the larger sex, and larger birds tend to have longer bills.

In the next section, we see that competition within a species is not always the main factor shaping the evolution of bird bill diversity.

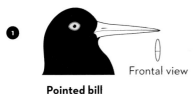

Frontal view

Pointed bill

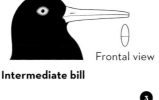

Frontal view

Intermediate bill

Bill lengths in oystercatchers

Among Eurasian oystercatchers, females tend to have pointier, longer bills (**1**) that are better adapted to pulling worms out of mud, while males tend to have bills better adapted for hammering open shellfish (**3**). Although the sexes differ on average, many individuals of both sexes have intermediate bills (**2**).

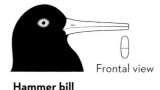

Frontal view

Hammer bill

CROSSBILL COEVOLUTION

In some birds, specialists are divided into variants that look different, sound different and have different food preferences. Very rarely, these divergent food choices can lead to the origin of new species. Crossbills are named for their bills, which cross at the tips and are perfect for prying apart cones to release the seeds from between the scales.

Red or common crossbills pose a tremendous classification challenge for biologists and birders. They occur in many subtly different forms, each seemingly best fitted to extricating the seeds from a particular conifer species. In the Rocky Mountains, hemlock, Douglas fir, ponderosa pine and lodgepole pine specialists are each fastest at feeding on their preferred conifer. The biologist Craig W. Benkman, who measured all these bird bills and ran little cone-processing contests in the late 1980s, even predicted the optimal shape for a fifth crossbill variant based on another common conifer, the Sitka spruce. When those birds were finally noticed over 10 years later, their bills indeed matched Benkman's predictions.

Crossbills from each clan will feed on any conifer species available, but prefer the cones they are best adapted to open. Birds that depend on unpredictable boom-bust food sources, such as red-breasted nuthatches, undergo localised population explosions called irruptions. If you have ever noticed a sudden glut of crossbills, that is an irruption. It happens when flocks of crossbills from any clan congregate to capitalise on a bumper cone crop.

Each clan has a unique flight call, best distinguished by translating sound recordings into pictures, called sonograms. These calls are important, not just for humans wishing to identify a crossbill variant, but also for the crossbills themselves. In order to find food, crossbills move about in noisy, nomadic flocks all year round. They choose their mates from within these feeding clans, and breed whenever there is enough of a cone crop, even if that means nesting in winter. There are at least eleven distinct clans in North America that remain somewhat genetically distinct by breeding only with crossbills that have similar flight calls. In Eurasia, there are at least twenty variants.

SETTLING DOWN TO SEDENTARY LIFE

Rather than the nomadic lifestyle of most crossbill clans, a handful of clans are more sedentary, thanks to a stable food supply. In a few isolated hills near the Rocky Mountains, crossbills have the luxury of settling in one spot, because the lodgepole pine cones they specialise on release their seeds slowly and steadily. This gradual loosening of the cones is an adaptation to fire: the pines do not release all their seeds unless the resin sticking the scales together starts to soften at high temperatures. During irruptions, nomadic clans will feed in the same place as resident crossbills. However, because of their distinct calls, the different clans seldom interbreed.

Sedentary crossbills in the South Hills of Idaho are engaged in an evolutionary arms race with lodgepole pines, because another major pine-cone predator, the red squirrel, has never colonised these hills. The lodgepole pines can afford to evolve cones that are fortified against crossbills, without also having to defend their seeds against squirrels. Crossbills have a harder time prying apart the scales on larger cones, whereas squirrels prefer large cones laden with seeds. So while most of the Rocky Mountain lodgepole pines have smaller cones with fewer seeds each, to discourage squirrels, the South Hills pine cones can afford to be big. This in turn has driven the resident South Hills crossbills to evolve much larger bills to open the larger cones, and so on, in an escalating coevolutionary arms race.

As a result of coevolution with their food, the South Hills clan is so distinct in size, sound, habits and genetics that it is now a distinct species, the cassia crossbill. However, climate change is leading to a rapid decline in this newly minted species, which suffered an 80 per cent decline in the population between 2003 and 2011. This is because warmer temperatures mimic the effects of fire, causing the cones to loosen their seeds prematurely, so removing the steady food supply the resident crossbills have come to depend on. Climate projections for the crossbills of Sweden and Scotland look similarly gloomy.

Western hemlock cone

Factors affecting crossbills and bill size

Different races of crossbill have subtly different bills adapted to prying apart the cones of different conifers. In the case of lodgepole pine specialists, the race that coexists with red squirrels (**4**) has a smaller bill than the race that has to pry open the larger cones made by lodgepole pines in the absence of squirrels (**5**).

Douglas fir cone

Ponderosa pine cone

Lodgepole pine cone: where squirrels live, the cones are smaller.

Lodgepole pine cone: where no squirrels live, the cones are larger.

TRADING BENEFITS

Not all relationships between birds and other species are as combative as the coevolutionary arms race between crossbills and the conifers they feed on. Cowbirds of the Americas got their name from always being found near cows. This is because they feed on the insects flushed from the grass by herds of grazing cattle and bison. Similarly, many unrelated South American birds are collectively referred to as antbirds, because they follow army ant swarms through the rainforest, capitalising on the collateral damage. These relationships are one-sided, as the birds are benefiting purely as a by-product of the behaviour of other species.

Below
Human honey hunters in Mozambique communicate with greater honeyguides to find wild beehives. In return, the birds are rewarded with beeswax, which they are specially adapted to digest.

Opposite
Bold, black-billed magpies and shy elk make the best partners.

By contrast, cooperation between humans and greater honeyguides in sub-Saharan Africa relies on a sophisticated form of inter-species communication. Honeyguides feed on beeswax, but have a hard time getting the wax without the help of a mammal, who calm the bees and break open the beehives. Human societies have long had a fondness for honey, but do not always know where the beehives are. Honeyguides make special calls to attract the attention of potential human collaborators. They can also communicate both distance and direction from a hive to the honey hunters.

In an especially elegant series of experiments in Mozambique, in 2016, the ornithologist Claire Spottiswoode showed that the resident honey hunters have a special call that they use to attract a hungry honeyguide. The honeyguides are three times more likely to respond to this call than to any other sound from the humans. This signal must be learned, because young greater honeyguides simply sit and stare, and different human societies have very different calls to attract their local honeyguide collaborators. For their part, the honeyguides triple the humans' chances of finding honey.

In mutualisms, individual personalities can make a difference. For instance, black-billed magpies in the Rocky Mountains enjoy picking ticks off large grazing mammals. However, it's not just any magpie and any elk that tend to spend time together. After painstakingly testing how outgoing and risk-prone individuals of both species were, biologists found that only the boldest magpies hang about with the shyest elk. Shy magpies are too nervous to risk sitting on the backs of elk, and the more aggressive elk will not tolerate being perched on, even if the magpie returns the favour by removing a debilitating load of ticks.

Special diets can require mutualisms with microbes. Vampire finches have evolved to specialise on the blood of other birds. This diet requires specialised gut bacteria, some the same as those found in the guts of vampire bats. On the more vegetarian end of the spectrum, hoatzins eat only leaves; they have their own contingent of symbiotic bacteria that break down plant fibre. Hoatzins have even evolved some of the same genetic changes as cows, to enable them to digest the bacteria.

COOPERATION OR EXPLOITATION?

Below

A fork-tailed drongo hovers over a foraging meerkat. The drongo 'cries wolf' to steal a free meal from other species, such as meerkats and pied babblers.

Cooperative relationships are vulnerable to cheating, and, even among birds, the connection between mutualism and parasitism is often blurred. For example, oxpeckers eat ticks off the backs of large herbivores on the African savannah. Most of the time, this benefits both parties, but experiments have also shown that when the ticks are in short supply, oxpeckers are not above helping themselves to blood directly from the tick wounds.

Fork-tailed drongos of the Kalahari Desert are often sentinels for a mixed-species group of foragers ranging from meerkats to pied babblers. This extra security allows everyone else to concentrate on feeding, rather than simultaneously scanning the skies for danger. While most of the species in these groups are social, and experts at excavating food from the ground, drongos perch solo on a high branch, from which they hawk for insects. Their short, straight bills are perfect for catching insects in midair, but poor tools for unearthing buried treasures such as fat beetle grubs. In contrast, pied babblers have thinner, curved bills that are better suited to probing and digging.

Rather than wasting time and effort excavating delicious delicacies, the drongos take advantage of the superior digging skills of other species. They do this by sounding a false alarm just when a pied babbler has unearthed a juicy grub. Drongos benefit from the deception by expanding their feeding niche beyond what their bills have evolved to cope with. Interestingly, individual drongos differ in how exploitative and parasitic they are towards the species they are ostensibly helping to guard.

Individual drongos also flexibly employ these deceptive tactics when they most need to, such as when it is too cold for most of their flying insect prey to be active. This strategically selective use of a false alarm call helps to prevent the other species from cottoning on and subsequently calling a drongo's bluff.

If you ever see a mixed-species flock, do not assume it is a perfectly harmonious situation in which all the species are benefiting from a neighbourhood-watch effect, while minimising conflict by sticking to a specialised feeding niche. All cooperatives are vulnerable to cheats – especially skilful ones like the fork-tailed drongos.

Above

Red-billed oxpeckers eat the ticks off herbivores, such as this impala, but can also follow that up with a chaser of blood from the wound.

PLANT POWER

Opposite

Several bird lineages, such as this oriental pied hornbill, have evolved to disperse plant seeds. Many tropical fig species, like this Banyan tree, rely on these fruit-eating birds to distribute their seeds far away from the parent tree.

Below

Cyrtosia septentrionalis is an East Asian orchid that produces uncharacteristically large, red fruit to attract birds like the brown-eared bulbul as seed dispersers.

Even plants can manipulate mutualisms. If you have ever been bothered by squirrels guzzling all the birdfeed, you may have used spicy red pepper to dissuade them. There are even products with names like 'Flaming Seed Sauce' one can buy to coat birdseed in fiery oil. The reason this works is that capsaicin, the compound that makes chillies taste hot to humans, is a pepper plant's way of selecting its seed dispersers. Birds lack the taste buds for capsaicin, so will feed on seeds anointed with hot sauce that would make a squirrel's tongue flame.

In the Sonoran Desert, chiltepin peppers, ancestors of domesticated capsicums, benefit from having their seeds dispersed by birds such as the northern mockingbird. This is because birds are more likely to deposit seeds in damper, more shaded areas where seedlings can thrive.

There are even orchids that have evolved sweet, fleshy red fruits to attract birds as seed dispersers. Most orchids have tiny seeds that are wind-dispersed. However, in the dank understory of a temperate Japanese rainforest, there lives an orchid that depends solely on fungi for nutrition – because there is not enough sunlight to photosynthesize – and relies on birds to disperse its seeds. In addition to producing attractive fruits, this orchid has even evolved an especially sturdy stem for feeding birds to perch on. Biologists spend hours watching these orchids. They have discovered that a common East Asian bird, the brown-eared bulbul, ate the most fruits, and also had a lot of orchid seeds in its droppings.

Other plants have coevolved more closely with their avian seed dispersers. Such specificity can become a conservation problem. Wekas are large, flightless rails (from family *Rallidae*) endemic to New Zealand. A handful of native plants, such as the hinau tree, depend on wekas to disperse their seeds. This was all well and good until human campers provided the wekas with more attractive alternatives: food leftovers. Biologists microchipped some of the hinau's large, unwieldy seeds in much the same manner one would a pet, and then waited to see how long each seed took to pass through the wekas' guts. In most fruit-eating birds, seeds take only a matter of minutes to pass from mouth to droppings, but large hinau seeds took an average of five days. Wild wekas would have travelled a useful distance by this time, but the effect of the campers' food waste has meant bad news for plants like the hinau, because their seedlings end up falling much nearer to the parent trees.

As we will see in the next section, there are also multiple instances of plants evolving to use birds as pollinators.

DOES A FLOWER BY ANY COLOUR TASTE AS SWEET?

In addition to helping plants place their offspring in an advantageous position by dispersing seeds, birds can be essential for successful plant sex. Plants such as monkeyflowers rely on pollinators to avoid breeding with an incompatible individual.

A hummingbird-pollinated monkeyflower has bright-red, narrow flowers with protruding reproductive parts that can reach a hummingbird's forehead, and loads of nectar as a bribe. By contrast, a bumblebee-pollinated species is pink and broad, with less nectar. By crossing these two closely related but very different plants, biologists could show that only a few genetic differences were enough to make a flower much more attractive to either birds or bees. This genetic flexibility means that monkeyflowers can quickly evolve into new species purely by attracting an exclusive pollinator. Hummingbirds can be taught to prefer other colours with enough of a reward difference, but have an innate preference for red.

BIRD NECTAR SPECIALISTS

- Sunbirds
- Hummingbirds
- Honeyeaters
- Lories

Most birds do not actually eat sweet things, and lack the ability to taste sweetness. My colleague Maude Baldwin asked how pollinators, such as sunbirds of the Old World (Africa, Asia and Europe), or hummingbirds of the Americas, evolved to find nectar a rewarding food. Taste buds work by having receptors that send signals to the brain in response to certain molecules. For instance, a sour taste is triggered by hydrogen ions released from any acid hitting a sour taste receptor. Similarly, the flavour *umami*, a Japanese word for the savoury taste of meat or soy sauce, and distinct from pure saltiness, is triggered by receptors for amino acids, which are the building blocks of meat. MSG is a synthetic amino acid.

Hummingbirds are most closely related to swifts, so how did they evolve the ability to even perceive sweetness, so that plants could attract them with a nectar reward? The genes encoding the *umami* receptor are evolutionarily old, and shared between a common mammal and bird ancestor. One of them also overlaps with a gene that enables most mammals to taste

sugar. However, this gene got lost somewhere along the dinosaur lineage leading to birds, so most birds taste *umami*, but not sugar. Remarkably, hummingbirds have independently re-evolved the ability to taste sweetness. Just six changes in the *umami* receptor sequence have converted it into a receptor for sugars in hummingbirds.

Above

Plants such as this monkeyflower rely on different flower structures and colours to attract different pollinators. These bright-red petals appeal to hummingbirds, whereas bees prefer broader, pinker monkeyflowers.

SMELLY FOOD

Having a sense of smell is a largely mammalian way of perceiving the world, which both humans and birds are generally thought to be pretty bad at. However, an experiment in 1985 showed that black-billed magpies found more buried raisins coated in cod-liver oil than unoiled raisins. Sniffing out buried food could be especially useful among corvids such as magpies and ravens that often pilfer from food caches. This is in spite of their olfactory bulbs – the part of the brain involved in processing scent – being disproportionately small compared to that of a rat or a dog.

By contrast, turkey vultures of the Americas are among the avian champions of scenting food. Looking at their brains reveals that even compared to another New World vulture, the black vulture, turkey vultures have the largest olfactory bulbs for their brain size compared to any other bird studied so far. Turkey vultures can find carrion in dense forests by smelling it. In contrast, black vultures and the completely unrelated Old World vultures rely more on sight to find carrion.

Turkey vultures also have enlarged nostrils. So do another group of birds with a highly developed sense of smell – the tubenoses – although in this case, the large nostrils could also help with sensing airspeed. These seabirds, including albatrosses, shearwaters and storm petrels, spend most of their lives soaring above the open ocean. They find food over vast distances by smelling a gas, DMS, which stands for dimethyl sulfide. This gas is released when microscopic animals feed on algae near the surface of the ocean, and has a smell reminiscent of seaweed or oysters.

Tubenose species that use DMS as a feeding cue are five times as likely to eat plastic than species that do not respond to DMS.

Penguins are also attracted to DMS, presumably because it's a reliable sign of good fishing. The downside of using DMS as a long-distance fishing beacon is that this same gas is also often produced by all the microorganisms that collect on the surface of discarded plastics in the ocean. Humans dump about eight million tons of plastic into the oceans every year, and biologists have puzzled over why an estimated 90 per cent of seabirds have swallowed plastic. In an experiment that put three common types of plastic out in the ocean for three weeks, biologists found that that was enough for all three plastics to accumulate surface concentrations of DMS to attract hungry tubenoses. However, there are also species, such as the Laysan albatross, that do not seem to use DMS as a cue, but that still die from eating too much plastic.

Above

New World vultures like these turkey vultures (called buzzards in America) have the same scavenging lifestyle as Old World vultures.

WINTER FOOD

People who live at high latitudes often claim to suffer more from the temptation of alcohol during long, dark depressing winters. It is also in these places that I frequently encounter the corpses of inebriated birds such as Bohemian waxwings, who have imbibed too liberally from a tree of fermenting mountain ash berries and crashed into a building.

In contrast to the nomadic life of Bohemian waxwings, which flock wherever they can find berries, the aptly named Townsend's solitaire adopts the strategy of guarding a small patch of berry bushes during the winter. Both sexes do this by singing to advertise their territories, each of which encompasses a few precious juniper bushes. It is quite unusual to hear birdsong in the winter, and doubly unusual to hear both sexes, so the pretty song of this American thrush is worth listening out for in the mountains of western North America.

Another strategy for staying fed in the winter is to cache food. Birds that do this, such as chickadees and tits, grow their hippocampus (the spatial memory part of the brain) every autumn. It is rather like having a removable external hard drive for storing data only when one needs the extra capacity. However, it is also possible for natural selection to enlarge the spatial hard drive in populations that need to store more food. In a study monitoring thousands of mountain chickadees over their lifetimes, biologists found that chickadees living higher up the mountains, and so experiencing harsher winters, had a larger hippocampus than those living at a more clement altitude. These high-altitude birds were also better at solving spatial IQ tests, such as learning which bird feeders were most rewarding.

Although most mountain chickadees lived for one-and-a-half years, a few reached the venerable age of seven, and these were no better than the younger adults at remembering where to find food. However, the juvenile birds that did better on the spatial memory tests were more likely to survive their first winter than their contemporaries with a poorer memory.

The main point is that small birds, with high metabolic rates, spend much of the winter in a state of economic threat, and have evolved a variety of behavioural strategies to cope with the cold, from being nomadic to evolving a seasonally enhanced spatial memory.

Opposite
Mountain chickadees feed on seeds. (Top) Note its fluffed feathers, called piloerection, for protection against the cold.

PLANNING FOR THE FUTURE

Some birds take food storage to another level of sophistication. Siberian and Canada jays amass communal larders for the winter by glueing bits of food under bark with a special sticky saliva. Real estate can be in such short supply that some fully grown individuals delay leaving the parental territory, or join a different family with an established territory. These hangers-on help the breeding pair to harvest and hoard food.

Similarly, some populations of acorn woodpeckers form groups that communally breed, hoard and guard acorn vaults. The woodpeckers hammer acorns into holes in tree bark, and the fit has to be just right. Prime acorn-banking trees are rare enough to be in high demand, and acorn woodpecker groups have to be large enough to defend them from neighbouring clans.

Florida scrub-jays are also group breeders. However, rather than storing food, they defend large territories as buffers against hard times. These birds need just the right sort of fire regime to produce the acorns and arthropods they eat, and a larger property is more likely to contain the right mix of recently burned and recovering oaks for the jays to have enough food over multiple years. However, such large holdings by established breeding pairs make it almost impossible for young jays to set up house on their own. The sons of Florida scrub-jays stay home to help their parents rear the next broods – not just because they cannot afford a place of their own, but also because they stand to inherit a piece of the family estate.

Perhaps the most impressive demonstration of mental prowess is found in western scrub-jays, which also hide food away for leaner times. In the lab, Nicky Clayton and her colleagues have shown that these birds are capable of mental time travel. They did this first by giving the jays two types of food to cache, and then varying when the jays had a chance to retrieve their stored valuables. Jays returning for recently stored food went for perishable wax worms, which they adore. In contrast, when returning after the preferred worms had passed their expiry date, the same birds went straight to the better-preserved, but less-delectable peanuts. In other words, scrub-jays do not just remember where they stored food. They also know what was put in each place, and when.

FEEDER FEEDBACKS

Bird feeders, especially in the US and UK, have had a dramatic impact on bird behaviour and evolution. A citizen science project run by the Cornell Lab of Ornithology, called Project Feeder Watch, has gathered enough data to show that some birds, such as the northern cardinal, have expanded their range northward, probably because of more supplemental feeding by humans. Even more striking are Anna's hummingbirds, which have expanded their range from California to Oregon, Washington and beyond British Columbia to Alaska.

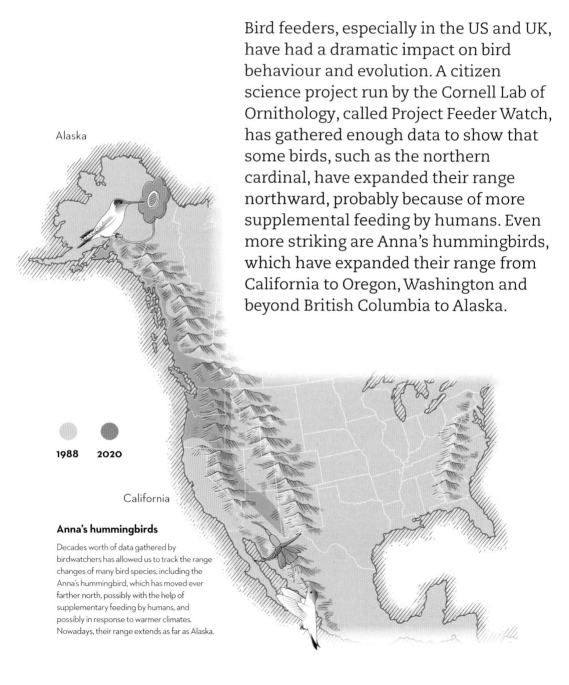

Alaska

1988   2020

California

Anna's hummingbirds

Decades worth of data gathered by birdwatchers has allowed us to track the range changes of many bird species, including the Anna's hummingbird, which has moved ever farther north, possibly with the help of supplementary feeding by humans, and possibly in response to warmer climates. Nowadays, their range extends as far as Alaska.

The same project has also revealed some interesting hierarchies that take place at feeders. Among 136 species, the heavier birds tend to dominate the lightweights. However, there are some exceptions, including orioles, and some warblers and hummingbirds, which do manage to dominate larger birds. Downy woodpeckers are especially aggressive for their diminutive size. Biologists now wonder if this is because they look so much like their distantly related but much larger hairy woodpecker, so other birds are duped into giving way to them. In contrast, doves, buntings and grosbeaks are less pushy than would be expected for their size.

The British famously love their birds, spending twice as much on birdseed than all of mainland Europe combined. At present, over 50 per cent of gardens in the UK have bird feeders. A study on bird populations in the UK from 1973–2012 showed that feeding birds increased in popularity over the 40-year span and, with increasing commercialisation, a greater diversity of birds flocked to feeders as the variety of food increased. The study found little change in species that do not use feeders, but birds as varied as goldfinches, great spotted woodpeckers, woodpigeons, long-tailed tits and blackcap warblers, all of which frequent feeders, have all increased in population size. Even bird predators, such as sparrowhawks and other *accipiters*, who capitalise on easy prey lured by bird feeders, have increased in population.

Fascinatingly, the British penchant for bird feeders could have selected for longer bills, enabling birds that frequent feeders to reach birdseed more easily. Great tits in the UK, for example, differ enough from those in mainland Europe to be given subspecies status – and one of the differences is a longer bill. To discover the cause of this difference, biologists compared the genes of great tits in the UK to those of shorter-billed relatives in the Netherlands. They narrowed down a gene that helps make collagen – a crucial component for growing a bill – that made a difference to bill length; significantly, Dutch and British great tits also have different versions of this gene.

The great tits of Wytham Woods, near Oxford, have been meticulously monitored by the university's biologists for decades. The evidence gathered shows that the tits in this population have evolved longer bills over 26 years, as bird feeders have become more popular. They also showed that within this British population, individuals with two copies of the long-billed flavour of the collagen gene were most likely to use bird feeders, while those with the short-billed genes spent less time at feeders. This suggests that every time we feed birds, we could be changing not just their diets, but also their behaviour – and ultimately, contributing to their evolution.

ECONOMIC FEEDING

Birds often forage in mixed-species flocks, also known to birders as bird waves, or bird parties, as they travel through a forest. This kind of group foraging occurs year-round in the tropics, but in more temperate climates is most marked in the winter. As groups get larger, birds must balance the costs of competition and attracting predators against the benefits of using social information (such as following the flock) to find food, and sharing the effort of being vigilant against danger.

The blue, great and marsh tits of Wytham Woods near Oxford are an especially good example of how birds that feed in flocks make economically sound decisions. Like many of their chickadee and tit relatives, these three species call to signal that they have found a good patch of food. These social signals transcend species boundaries, and make finding food in winter much more efficient than going solo.

A bird party in a Malaysian rainforest

Mixed-species flocks often feed together, in what is commonly known as a bird wave, or bird party. Everyone benefits from having more eyes and ears on the alert for danger, but the greater racket-tailed drongo (**3**) is the most likely to act as a sentinel, possibly because these birds already perch in the open to catch insect prey on the fly. The drongo is joined here by a chestnut-crowned warbler (**1**), a green magpie (**2**), a blue nuthatch (**4**), and a sultan tit (**5**).

Biologists recording these mixed-species flocks found that birds were the most vocal early in the morning, when flocks were still small. Broadcasters got more group members to watch against predators, while new recruits benefited by tapping into the local food news network.

However, making a food call also has its risks because it alerts predators. By playing back food calls later in the day, biologists showed that birds were still responsive to social information about food. This suggests that the mixed flocks got quieter not because food calls were no longer useful, but because the costs of being noisy outweighed the benefits, as the day wore on and groups got bigger. House sparrows, willow tits and Carolina chickadees all make fewer food calls as feeding flocks get larger, possibly because the risk of attracting predators increases as the group gets bigger. When you next see a flock of birds feeding, take note of how noisy they are and consider what they might be communicating, and why.

HANDY BIRDS

One of my favourite birds is Vultchy. He lives on the outskirts of Nairobi, with some very talented friends who are artists and biologists. Vultchy enjoyed sharing my porridge in the mornings, which is surprising because Egyptian vultures are more famous as egg eaters. In 1969, Jane Goodall published a letter in *Nature* reporting the use of stone tools not by chimpanzees, but by Vultchy and his friends. Egyptian vultures are adept at using stones to crack ostrich eggshells. Another raptor, the black-breasted buzzard of Australia, has independently figured out how to use stones to crack the shells of emu eggs.

Above

A woodpecker finch, one of Darwin's finches, using a stick tool to probe for grubs in the Galápagos.

Opposite

Egyptian vultures are among the first birds documented as tool users. This one is using a stone to break open an ostrich egg.

There are precious few birds that habitually use tools in the wild. The aptly named woodpecker finch in the Galápagos does often forage like a woodpecker. However, it also breaks off sharp twigs or cactus spines with which to spear and pry grubs out of cracks in tree bark. This tool-using behaviour is something all young woodpecker finches seem to learn on their own. There is no evidence that being exposed to tool-using adults speeds up the learning process.

In contrast, New Caledonian crows are arguably the most sophisticated handy birds, and there is enough social learning for different tool cultures to develop. One of the materials these crows use in the wild are long, narrow, pandanus leaves that the crows tear and shape with their bills. Crows in different parts of the island of New Caledonia have at least three distinct patterns of tool that they make, which have no clear relationship to their environment.

Biologists have puzzled over how to be sure that these regional leaf-tearing styles are the product of culture, because the crows do not copy each other's movements directly. A recent experiment shows that New Caledonian crows visualise the desired end result as they watch what their compatriots make. They then replicate the design without having to imitate the process of manufacture.

ISLAND-INSPIRED INGENUITY

The capacity New Caledonian crows have for conceptualising solutions is most famously demonstrated by Betty; she worked out how to bend a straight wire into a hook to extricate a bucket in an Oxford lab. A popular video shows her pulling this off without any trial and error, and then having her mate, Abel, swoop in to steal the spoils. More recently, experiments have presented naive crows with short lengths of stick that fit together like Lego pieces. The results show that the crows simply look at the bits, realise they need a much longer tool to reach the food, and literally put the pieces together.

Scientists know of a second crow, also from a Pacific island, that also makes tools. Hawaiian crows are not quite as innovative as New Caledonian crows, but over 90 per cent of birds in a captive breeding programme modified sticks into tools when given the chance. Like New Caledonian crows, they have unusually straight bills and very mobile eyes, which biologists speculate are like the opposable thumbs of the corvid world. As Hawaiian and New Caledonian crows are very distantly related, and none of the species evolutionarily closer to either of them makes tools, it is thought that this mental capacity evolved independently in the two island species. Why, you may ask, are the crows from Pacific islands so extremely ingenious? Perhaps freedom from predators, and proportionately high caloric value allows such inventiveness to flourish.

Nevertheless, there are plenty of other birds that are clever at manipulating objects. For decades, animal behaviour texts were filled with stories of blue tits learning to peck away the foil and skim cream off the tops of milk bottles left on the doorstep. More recently, there are reports of sulfur-crested cockatoos opening rubbish bins in Sydney, and Japanese crows using vehicles to crack nuts on traffic crossings so they can take their time eating when pedestrians are crossing. So, even if you do not live on a Pacific island, it is worth observing the birds around you feeding. You might just notice them discovering a new trick.

Opposite

New Caledonian crows make a variety of tools in the wild, some of which appear to follow region-specific styles in different parts of the island of New Caledonia.

A SOCIAL BIRD

Right

Like other species in the genus, these purple-crowned fairywrens are cooperative breeders, in which a dominant pair is assisted by 'helpers at the nest', who may or may not also be breeding on the sly.

PHYSICAL AND SOCIAL INTELLIGENCE

The corvids are generally acknowledged to be among the brainiest of bird families. We know from anecdotes and experiments that jays, jackdaws, magpies, crows, nutcrackers, rooks and ravens do all sorts of clever things, from displacing water to deceiving each other, from playing to mourning their dead.

There are two explanations, not mutually exclusive, for why corvids are so clever. These are strikingly similar to hypotheses for human intelligence. One is that the process of finding and storing food selects for the ability to remember details of where, when and what food is stored, and to make tools. It also helps one to plan ahead. Ravens exercise self-control by flexibly changing their feeding behaviour in anticipation of a different reward in the future.

Another explanation is that social living can sometimes lead to multiple mental adaptations for dealing with other individuals. Skills include the ability to follow another bird's gaze, recall where another bird stored food, and deceive potential thieves by anticipating their actions.

Western scrub-jays, Eurasian jays and ravens all hide their food behind an opaque barrier when in the presence of another bird – suggesting that they know a potential thief is watching them. In addition, when ravens and scrub-jays know they have been observed, they return to move the food around, hiding it from a potential thief. They do this if the observer was a dominant bird, but not if they were observed by their partner or a subordinate individual. In contrast, naive scrub-jays that have never stolen food are less wary, and fail to move their food around to confuse thieves.

Corvids are not the only highly intelligent and social birds. Several Old World babbler species are cooperative breeders, and show signs of having a high emotional intelligence (EQ) reminiscent of primates. In 2019, biologists established that wild Arabian babblers use what psychologists call 'joint attention'. This means that one bird can direct another's attention to something of mutual interest.

Arabian babblers employ joint attention in at least two very different contexts. One is called 'babbler walk', used by adult caretakers to signal to a fledgling that they urgently need it to go somewhere. A caretaker signals by waving its wings and calling to the fledgling, then turning away and hopping towards the intended destination. Fledglings do not usually trail after their caregivers, but if solicited by a 'babbler walk' they will usually respond by following, or sometimes beating their caregiver to the intended destination. A caretaker with a recalcitrant youngster will keep looking behind to check, then hop back, repeat the wing waving, and hop away again until the dawdler follows. The other use of joint attention involves arranging a rendezvous between adults.

Food thieves

Eurasian jays, like other social and intelligent crow relatives, are more likely to go behind an opaque barrier to hide food if they know they are being watched.

SABOTAGING RELATIONSHIPS

We are all familiar with power couples. Individual ravens remember and monitor the relationships of others, sometimes even sabotaging budding couples to reduce future competition.

Unlike most songbirds, ravens, rooks and jackdaws form monogamous pair bonds for life. Couples that have been together for a long time reinforce their pair bonds with increasingly reciprocal behaviours, such as preening each other (allopreening), playing or engaging in dominance displays as a couple. Established couples will also support each other in conflicts with a third party, consoling a mate who has lost a fight, and celebrating with a victorious mate. Rather like humans holding hands, these corvid couples engage in bill twining after their partner has had a social altercation. They also preen and feed each other as ways to offer moral support to a stressed partner.

Ravens live in hierarchical societies where one's position in the social pecking order depends on a stable, long-term breeding relationship. Power and position are directly related to wealth in the form of a territory and access to food.

It can take years for a raven couple to become an established breeding pair with their own territory, sharing power in the top ranks of raven society. Just below territorial pairs in rank are close couples without a breeding territory. Below them are casual couples, at the most vulnerable stage of pair bonding. At the very bottom are single ravens, which are never seen engaging in pair-bonding behaviours repeatedly with another individual. How long the bouts of bonding behaviours are, and how equal the give and take may be, is a reliable measure of a couple's stability and status, both for biologists and for ravens.

DOMINANCE RANKINGS IN RAVEN SOCIETY

- Breeding pairs with an established territory
- Close couples without a breeding territory
- Casual couples
- Non-bonded individuals

Dominant couples, particularly those with established territories, intervene when less stable couples are actively bonding. Casual couples are targeted most, while no one bothers to sabotage singles engaging in a first foray with a new partner. About half the interventions successfully split up a couple. Sometimes the couple manages to fight back and repel the saboteur, while other times the outcome is unresolved, with all the birds staying or flying away together.

There is no immediate benefit to intervening in the billing and cooing of another established couple, and the saboteur risks a

Above

Sabotaging a budding relationship can involve acts of aggression that escalate into a fight. Aggressive disruptions are about twice as likely to split up a casual couple than when an intervener simply inserts itself between an established couple engaged in a bond-strengthening behaviour.

costly fight. However, the fact that casual couples at the most vulnerable and easily disrupted stage of pair bonding are the main targets suggests that ravens strategically sabotage budding alliances to retain their own social position. This will bring them greater benefits in the longer term.

SHADES OF SOCIALITY

Even though all corvids are social, there is a great deal of variation, both across and within species in group size and structure, which mirrors the variation seen in birds more generally. Unlike the highly territorial raven couples that occupy a substantial acreage of their own, colonial corvids such as rooks, jackdaws and pinyon jays live in more fluid, fission-fusion societies akin to primates such as chimpanzees.

In contrast, other corvids, such as Florida scrub-jays, have an entirely different social structure. Rather than clustering a lot of nuclear families into close quarters in a colony, cooperative breeders live and breed in groups where at least one adult is a non-reproducing helper at the nest. Biologists think the two main social structures – colonial living and cooperative breeding – arose independently via different paths. Colonial species such as rooks are akin to human nuclear families living in close quarters on a city block. Cooperative breeders such as Florida scrub-jays, Alpine choughs and Mexican jays resemble large, extended families that live on a vast estate. They are accompanied by nonbreeding subordinates, who assist the dominant pair with childcare, property maintenance and defence.

Other cooperative breeders vary their social system according to the harshness of the environment. Western scrub-jays in the lush Central Valley of California live in pairs, together with the odd offspring that has not yet found a territory of its own. Yet in parts of Mexico with a harsher climate, the same species live in extended families with multiple helpers, just as their relatives, the cooperatively breeding Mexican jays, do.

Similarly, carrion crows in arid parts of Spain breed cooperatively, but those in Switzerland do not. These differences are due to necessity born of environmental hardship, as crows switched as eggs adopt the social system of their foster parents.

There is also variation at the other extreme of social complexity. Black-billed magpie pairs in Europe form evenly spaced territories, within which they forage and breed year-round. In contrast, North American populations have pairs that nest and feed closer together, instead of in exclusive territories, and only use their territories during the breeding season. The yellow-billed magpie, a distinct species found only in California, is even more gregarious, with colonial pairs that live in flocks all through the year.

Below

A dominant pair of Florida scrub-jays is assisted by other adults (mostly adult sons), who stay at home to help with childcare and territory defence.

SOCIAL MEMORY

Individual recognition is especially complicated for those living in a society. Just think of how many people you know to differing degrees, and in different contexts. American crows and Australian magpies in the same group learn to sound more similar to each other than to members of other groups. Rival groups of cooperative breeders are not unlike human clans, employing distinctive insignia or tartan in vocal form. Mexican jays use their voices to distinguish individuals from different groups. In experiments where biologists play back voice recordings of individual birds, Mexican jays respond faster and more often to the voices of neighbouring group members than to the sounds from their own group. This is similar to the Montagues immediately investigating a potential intrusion or taunt by the Capulets.

Surprisingly, a large brain is not always needed to live in a socially complex society. Vulturine guineafowl, which, like most members of the chicken order, are not known for their intelligence, move about in flocks as large as sixty or more individuals. I have frequently woken up to the sound of a flock moving through Mpala Research Centre, just north of Mount Kenya, and assumed these birds were just loosely associated into a band that liked to congregate near houses for food.

In 2019, researchers from a Max Planck institute monitored individual guineafowl. To everyone's astonishment, they found that within each loose band, vulturine guineafowl actually feed and sleep in consistent, multitiered social groupings. At the core of these flocks are stable, nuclear groups of a few breeding pairs with hangers-on. Over the course of months, each group consistently flocks with their preferred groups, rather like human families gathering regularly for meals. Biologists used to think that only large-brained animals, such as corvids, babblers or primates, were capable of these complex, tiered societies. However, the vulturine guineafowl show us that even backyard fowl can keep track of multiple individuals – not just in their core group, but in other groups, too.

Opposite

Vulturine guineafowl of northeast Africa are surprisingly adept at keeping track of social relationships between multiple, loosely associated family groups. Adults have red eyes, while immature birds have black eyes.

EGGSPENSES

As with relationships between species (see chapter 1), a fine line exists between mutually beneficial and exploitative relationships within species. The anis of Central America are also group breeders – with a few twists. Unlike the colonial corvids, several ani pairs share a nest. Unlike cooperatively breeding scrub-jay or carrion crow families, there are no nonbreeding helpers at the communal nest.

Guira cuckoos, greater, groove-billed and smooth-billed anis are all members of a New World cuckoo subfamily with very similar societies. All four species are joint nesters, and females do their best to sabotage each other's breeding attempts so as to make more room for their own eggs. However, having too many eggs in a clutch reduces incubation efficiency, and a smaller proportion hatch, which is bad for everyone.

Although the outcome of this killing is greater laying synchrony, and a more egalitarian distribution of eggs across members, it comes at great individual cost. In particular, females that are first to lay lose the most eggs to other females. This is because females seem to follow the rule of thumb 'stop chucking out eggs once I've started laying', to avoid killing their own young by mistake. Females that start laying last, can spend the most time sabotaging others without that danger.

Egg laying is very expensive for anis. Each egg comprises 15–17 per cent of a female's body weight, and laying more eggs to compensate for egg loss can shorten a female's lifespan. Female smooth-billed anis nesting alone, or with one other female, lay an average of five to six eggs a season,

but early layers in large groups of four to five females have to lay as many as thirteen eggs to compensate for the greater number of eggs lost.

Smooth-billed anis in Puerto Rico go a step further by tossing eggs out of nests. They also bury eggs, which biologists can count only by waiting until the birds have finished breeding, then dismantling the large nests to search for eggs embedded in the matrix of sticks. There are also hints that, unlike egg tossing, burying is performed by both sexes.

The larger the groups, the more conflict and waste occurs, because females have to lay disproportionately more eggs to compensate for the greater number buried. The mean number of eggs in a smooth-billed ani nest soars by over nine eggs for each additional female in a group. While a female nesting alone lays up to seven eggs, groups of four to five females can have as many as fifty-five eggs in the communal nest. The upshot is fewer eggs hatched per female, and a higher rate of chick death in larger groups.

Biologists have measured stress hormones from the tail feathers of smooth-billed anis, revealing that members of larger groups had the highest levels of stress hormones during the breeding season.

LESS EQUALITY BETWEEN KIN

Breeding is more evenly distributed across group members in some species than in others, and this leads to differing amounts of within-group conflict.

DNA fingerprinting shows that smooth-billed anis differ from the other two ani species in having some relatives in the same breeding group. They are also the least egalitarian among males, with the dominant male siring significantly more chicks than other males – possibly through extra-pair matings within the group. Smooth-billed ani males are also the only ones seen joining their mates in tossing and burying eggs. This state of affairs is unsurprising, since the dominant male typically performs the most dangerous and labour-intensive incubation duties of the entire group by taking every single night shift.

In Costa Rica, dominant groove-billed ani males pay a high price for taking the night shift – the time in which they are most likely to be picked off by predators. Unlike their smooth-billed relatives, these dominant males sire a smaller majority of the chicks, and have to place most of their bets on their mate's status. Two-thirds of the dominant males are paired with the female that lays last, and so gets to throw out the most eggs and dominate the clutch. The remaining one-third somehow bets on the wrong female and mates with the first to lay, usually because the couple were recent immigrants to the group and the female is lower in the pecking order than her mate. In these cases, the female often attempts to force a re-nesting, so as to switch up the laying order. If that fails, the female or both members of the pair leaves the group.

At other times, stability of the groove-billed ani group can be achieved when the most senior female in the group switches

Opposite

A group of groove-billed anis shows a juvenile on the left, a female in the middle and an adult male to her right. Like other ani species, breeding pairs share a nest.

mates, pairing instead with the dominant immigrant male. While male dominance is largely a function of age, females establish their laying order through dominance displays, which involved lining up broadside to present their imposing profiles and heavy bills, while uttering a loud 'conk'. The last female to lay is usually the one with the tallest bill.

Greater anis from Panama are the most egalitarian of the three species – perhaps because they are the least likely to share a genetic investment with other group members. As a result, breeding members are the least likely to tolerate reproductive inequalities. In contrast to the other ani species, greater ani females almost always switch up the laying order across nesting attempts. Dominant male greater anis have no more extra-pair young than other males. In contrast to the stressful and wasteful habits of smooth-billed anis, individual greater anis raise more chicks in larger groups, because having more guards makes for less nest predation. Although the dominant male still performs nightly incubation for the whole group, he does not seem to face the same risks as do groove-billed anis.

FROM COOPERATION TO CONFLICT

At what point does joint nesting become brood parasitism, in which breeders could simply offload their eggs on caring individuals and escape all the work of parental care? Unlike some of their more famous relatives, such as the common cuckoo, anis do not specialise in laying their eggs in the nests of other species. However, some greater anis do parasitize each other.

When biologist Christina Riehl swabbed the surfaces of greater ani eggs to perform a maternity test, she found that 40 per cent of nests contained eggs laid by females from other groups. These parasitic pairs would stand to gain extra offspring without having to contribute to rearing them. However, greater anis have a clever way to identify parasitic eggs and toss them out of the nest before they can compromise the hatching success of the group's own eggs.

Greater ani eggs are covered in a chalky white coating. This wears off a few days after incubation begins to reveal a beautiful blue shell. Parasites from outside the group tend to lay their eggs a few days after incubation begins, leading to a mismatched white egg clearly conspicuous among a clutch of blue eggs. An elegant experiment, in which Riehl swapped eggs between nests, revealed that greater anis always tossed out introduced eggs that did not match their group clutch. This happened for the realistic case, in which white eggs added to a blue clutch were

rejected, but anis also rejected blue eggs added to a fresh clutch of white eggs. Indeed, the hosts rejected some of their own eggs if they were mismatched.

SINGING FOR SOLIDARITY

On the more cooperative side of social interaction, greater anis engage in group choruses, perhaps to synchronise egg laying. Female greater anis that have nested together in the past tend to synchronise egg laying faster than with newcomers. This increased laying synchrony leads to higher fledging success for all by reducing the cost of egg laying – this is because females have to go through fewer bouts of wasteful egg tossing. All members benefit from being in the same group for a few years, rather than switching to a new group. Incubation, which is shared by all breeding adults, even if one male performs the night shift, begins when the last or penultimate egg is laid.

SOCIAL SPECIALISTS

Ultimately, it suits individual greater ani females to specialise in either cooperating or cheating, rather than spreading their eggs too thinly between both strategies. This is because greater ani nests often fail due to predation, so females that switch between re-laying in the communal nest and dumping eggs in the nests of others have fewer offspring than those who specialise.

MUCH ADO ABOUT SOMETHING

Birdsong is a signal both to mates and competitors. As the following chapters discuss courtship and breeding, we will focus here on song as a territorial signal. Male eastern towhees are highly territorial, and can remember their neighbours' songs for years. If biologists play a recording of an old neighbour's song, a male is not too bothered, but he immediately goes to investigate if he hears a recording of an unfamiliar male next door. Similarly, male black-capped chickadees also engage in singing contests with neighbours, a behaviour known as countersinging.

In European robins (namesake of the American robin, which also has a red breast), males judge competitor quality by the complexity of their territorial songs. However, a study on noise pollution in Northern Ireland shows how urban noise interferes with the robins' abilities to hear the finer points in each other's songs, and compromises their judgement of potential rivals.

Singing is not restricted to males. Female northern cardinals also sing to claim their territories, and respond to other females by countersinging, whereas males typically start a fight with a male intruder. Interestingly, young female cardinals learn to sing three times faster than males, but males end up with more consistent songs. Female birds are more likely to sing in species or populations where female competition for mates or

Right

A male eastern towhee sings to claim his breeding territory. These birds know their long-term neighbours by remembering their songs.

Opposite

Contrary to the widespread assumption that only male songbirds sing, female cardinals sing to advertise their territories.

territories is high, or when they live year-round on a territory. Female birdsong may have been somewhat underappreciated in the past, but it is also possible that it is on the rise in some populations. In the 1980s, some juncos in San Diego stopped migrating, and have been carefully monitored by biologists ever since. Ellen Ketterson and her colleagues were surprised to notice some of these sedentary females singing, and so set up an experiment to see what started the female singers off. A handful of female juncos responded to a female intruder by singing, but such behaviour has yet to be recorded in other dark-eyed juncos.

Both sexes probably sang in the common ancestor of modern songbirds, after which females of many temperate species lost the ability to sing. Formally, the oscines (Latin for songbird) are the group of passerines, or 'perching birds' with complex voice boxes. The suboscines, by contrast, have a simpler voice box, and most of these species live in the tropics. Most song-learning studies focus on the oscines, because very few suboscines need experience or a song tutor to hone the song with which they were born.

MUSIC MECHANICS

Opposite

The silvereye is one of many species of white-eye, an Old World songbird group in which most species have distinctive white rings around their eyes.

Humans across cultures have long been fascinated by birdsong. Wolfgang Amadeus Mozart (1756–91) incorporated parts of his pet starling's songs in his compositions, while modern advances in neuroscience involve studying the parallels between birdsong and human language.

Island-hopping silvereyes

Starting in the 1830s, silvereyes have island-hopped from mainland Australia and a few offshore islands such as Heron and Lord Howe Island to Tasmania, before colonising New Zealand and other nearby islands. Their local song dialects have evolved with each move, so that the silvereyes on Lord Howe and Norfolk Island have songs that are separated by millenia of cultural evolution, and sound much more different than one would expect from their geographic proximity.

Heron Island
~4,000 years ago

AUSTRALIA   Brisbane

Norfolk Island

1904

Lord Howe Island
> 100,000 years ago

North Island

NEW ZEALAND

1865

1856

TASMANIA

1830

1856

South Island

Chatham Island

True songbirds typically learn in two stages. Young birds first listen to tutors (typically of their own species), using the examples to refine a rough innate template. Then they start babbling, until their vocalisations match their mental version. This process is analogous to how human infants learn language. They may have the capacity to grasp grammatical structures, but without listening to human speech at the right stage of development, and babbling to winnow down the extraneous sounds, they will not learn to speak. This social learning also explains why both humans and songbirds have regional dialects.

Birdsong can evolve in response to both cultural and natural selection. Silvereyes belong to a large group of small, greenish-grey Old World songbirds, highly prized in Asia as caged songbirds. Silvereyes have colonised a whole series of islands off the coast of Australia, including Tasmania and New Zealand in the 1830s, and eventually Norfolk Island in 1904. In contrast, silvereyes have been on islands much closer to Australia for thousands of years. Just as with many Polynesian human languages, in which the youngest societies tend to have the fewest consonants (think of how long words tend to be in Hawaiian), the youngest silvereye colonies tend to have songs with the fewest syllables. It is almost as though vocal complexity is slowly lost over successive migrations to new islands. However, the story is not quite so simple, because eventually the silvereyes on islands seem to regain song complexity, presumably through a process of cultural innovation.

SONGS SHAPED BY THE ENVIRONMENT

Furthermore, the older silvereye colonies reinvent songs that better match the acoustic properties of environments. Like many animals that communicate vocally, birds tend to have song structures that are most easily heard in a particular environment. Just as a siren cuts through urban noise with ease, birds in cities, or in more closed environments such as forests, tend to sing higher-pitched, simpler songs. Lower sounds do not carry as far through air, and complicated songs would get distorted from bouncing off myriad obstacles in a dense city or forest. Part of this song variation is learned, and part of it seems to be genetically inherited, changing more slowly through natural selection. Human language could also be shaped by the acoustic environment, because languages in dry, open landscapes have more consonants than those spoken in the hot, humid and dense tropical forests, where the nuances of different consonants do not carry as far.

SINGING UNDER STRESS

Most of what we know about the mechanics of song production comes from zebra finches. These birds are native to Australia, but are now found in laboratories worldwide. We know a lot about which bits of the brain are involved in memorising and producing songs. We also know that these brain regions tend to be more pronounced in males during the breeding season, and to shrink seasonally if not needed.

More recently, biologists have also found that male zebra finches that were underfed as nestlings have a harder time memorising their tutor's songs; females who were underfed appear less discriminating when they pick a mate as an adult. For males, this memory impairment extends into adulthood. One of the reasons for this is that the stress of being underfed actually causes hundreds of genes involved in song learning to be turned down. The young bird's brain is being programmed by the early stress to keep the song-learning switches partially 'off'. Similarly, because zebra finches are highly gregarious in the wild, a single night of solitary confinement is enough to dial down hundreds of genes involved in social communication in an adult zebra finch.

Opposite

Zebra finch males can be distinguished from females by their orange cheeks.

DUETS AND CHORUSES

In some birds, singing is a social activity involving both sexes in duets or choruses. There is some evidence that in species such as rufous-and-white wrens, pairs are using duets to keep tabs on their mates. However in most cases, duets and choruses serve to advertise a pair or group territory.

Red-crowned cranes of Japan often gather in large feeding flocks in winter, but breeding pairs accompanied by their young adult offspring are more likely to duet than 'childless couples'. Duets begin and end with a 'threat walk', in which both partners walk with exaggeratedly slow strides, their necks stretched vertically. In addition, the amount of duetting increases as more cranes join a feeding flock. This suggests that the cranes are using duets to jointly defend food for their families. Another function of duets is between parents coordinating childcare (see chapter 4).

Stripe-headed sparrows are highly territorial cooperative breeders. The females use song more for territorial defence than do males, who mostly use it in courtship. When biologists played recorded songs of both or either sex, females always responded to intruders before males, and responded more strongly to pair and same-sex intruders than did males. New Zealand bellbirds and fairywrens also have territorial females that sing both solo and in duets with their mates, to defend their turf.

Females take the lead in most duetting birds, but there are a couple of exceptions. Rufous horneros (red ovenbirds) of South America produce highly structured duets, but these are initiated and led by the male. In eastern and southern Africa, a garrulous little bird called the white-browed sparrow-weaver also sings duets in which male and female alternate with exacting precision.

Curious about how the birds achieve this coordination, biologists placed tiny radio transmitters on free-ranging sparrow-weavers. Transmitters on their backs recorded the birds' songs, while brain implants recorded every pulse from the region responsible for generating rhythm in a songbird brain. Both voice and brain recordings showed that the males led in each duet. Once his partner joined in, his mental metronome would slow down, and both birds would sing in synch to a slower internal rhythm. These duets sometimes extend to choruses, as white-browed sparrow-weavers are cooperative breeders.

Opposite

A pair of European rollers sing a duet.

TERRITORIAL DEFENCE

Birds do not just use sound to advertise their territorial boundaries. They also use visual signals, and if that fails, may resort to armed combat. The more evenly matched competitors are, the more fights are likely to escalate to determine the winner.

Blue tit males with crests that reflect more UV light are of higher quality, both as fighters and parents. As these birds see in the UV light, other blue tits respond to the brighter crests as signals of superior quality. Male blue tits are much more aggressive towards others with similarly bright crests than to males with artificially dulled crests; the latter pose less of a threat and are somewhat beneath their notice. Female blue tits that mate with these sexier males produce more sons.

Competing for densely packed territories in cities has had an interesting effect on bird personalities. Personality traits such as aggression to intruders of the same species or humans, and boldness, measured by the willingness to experiment and explore, tend to come in a package for many species. House sparrows and American song sparrows (one of many examples of completely unrelated birds sharing a common name because they look superficially similar) are bolder in cities, and also more behaviourally flexible. What this means is that traits defining a personality syndrome often come uncoupled in urban populations. Great tits in Barcelona, Spain also showed the same pattern of having a more varied mixture of personality traits, and being much bolder than their country cousins. In contrast, a study looking at great tits in the UK found that the urban birds, although bolder, retained consistent personalities.

Like many experiments on bird territoriality, these involved playing a competitor's songs, and recording how resident great tits responded. Bolder birds were more aggressive, approaching the speakers sooner, closer, and more often. Whenever I see birdwatchers using a recording to lure in a much-desired specimen, I wonder if this is a bold individual, who might get an ego boost from quickly dispatching the intruder (assuming the human stops the playback before long), or a shy bird that will be intimidated and extra stressed for the rest of the day.

Black coucals, a species of African cuckoo, are an exception both to the assumption

that males are more likely to sing to defend their territories, and also to the idea that singing is an oscine speciality. Female black coucals are larger than males, and compete for territories and a male harem with their voices. Larger females tend to have deeper voices, signalling their superior size. However, when challenged by recorded playbacks, individual females will artificially deepen their voices to sound more intimidating.

SWEET NOTHINGS

Birds produce a lot of sounds in addition to singing. They have multiple ways to strengthen and affirm a pair bond, including nonvocal attentions.

Green-rumped parrotlets use contact calls to communicate from a distance, even when out of sight. These birds breed in such high densities that ten socially monogamous pairs may nest within earshot of one another. Females, which incubate the eggs in cavities, are often unable to see their mates.

Captive cockatiel juveniles share food and preen each other – a behaviour called allopreening. Young jackdaws do the same, largely as a prelude to courtship. In contrast, young cockatiels are just as likely to perform these favours for either sex. They are also more likely to share food with their siblings, and with unrelated birds that reciprocate preening. However, the number of eating and preening partners dwindles as the cockatiels grow up.

Ravens present their mates with nonedible objects such as moss, pebbles or sticks. Although ravens do not coo, these symbolic offerings elicit a bout of affection from the recipient in the form of billing, or a bout of 'joint object manipulation'.

Arabian babblers (see page 50) use objects such as eggshell fragments, twigs or leaves, to signal to a prospective sexual partner that they want to retreat to a private spot to copulate. In most cases, males present the tokens, and females seem to know full well what is intended. If interested, the female bird sticks her tail in the air and presents her rear end to her partner – a fairly universal bird signal for saying she is ready. At times, females also take the lead to the rendezvous spot, or reciprocate with tokens of their own.

TOKENS OF AFFECTION

Some birds present tokens of affection such as:

- Eggshells
- Flower petals and food
- Leaves and moss
- Pebbles and sticks

While most birds are not at all shy about copulating in public, Arabian babblers actively seek out a thick bush for privacy. This apparent bashfulness is not restricted to furtive copulations by subordinate group members; even the dominant pair conducts their object presentations in private. One can only speculate why because this research is so recent. Some biologists think that the dominant breeding pair could gain by being discrete; it may help to keep subordinates aspiring to breed hopeful – so willing to assist with cooperative childcare and other duties.

Above

Green-rumped parrotlets form strong pair bonds, and rarely switch mates or engage in extra-pair copulations.

RESOURCE DISTRIBUTIONS

How food and other resources are distributed can dictate social systems, both within and between species. When resources are clumped, like the reeds that red-winged blackbirds in North America nest among, a male's best bet is to establish an attractive property, in the manner of the landed gentry in Jane Austen's books. It is a truth universally acknowledged among biologists (and birds) that a single blackbird in possession of a territory must be in want of a wife.

Although open polygyny would not be proper in many human societies, red-winged blackbird females will prefer to be the second, or even third wife of a male with an outstanding territory, than to monopolise the parental help and attention of a male offering less material wealth. Biologists have demonstrated this general explanation for red-winged blackbird mating systems through a series of experiments in Canada. They removed females from some polygynous groups to create monogamous couples, then counted the breeding success of the different families. The scientists found that monogamous females raised more offspring than those in polygynous relationships, because they suffered less predation, while the males helped more with feeding chicks.

In another experiment, biologists removed females from territories of equal quality. New females chose to settle with single males in preference to those already paired up. All else being equal, female blackbirds do better in monogamous than polygynous relationships. However, in thirteen pairs out of sixteen, females chose to be the second wife on an improved territory, where biologists added nesting platforms over water, over monogamy in an inferior territory, where biologists had removed all the prime nesting sites safe from predators. Reed warbler and red-winged blackbird females that become second wives of a male on a good territory could be making similar economic choices to monogamous females on poor territories.

Yellow-headed blackbirds, by contrast, feed far away from their nesting site, and males offer no help with childcare, so females are content to nest close together and share mates. Another New World blackbird, the yellow-rumped cacique, also has a polygynous mating system because paternal care is unnecessary. There is thus a negligible cost to females sharing a husband, and a benefit to nesting close together to ward off predators.

When breeding resources are so scarce that both sexes are limited by the number of eggs females produce, polyandry – in which a single female breeds with multiple males – occurs. The breeding habitat for northern jacanas in Costa Rica is very scarce; males carve out tiny territories in which they do most of the childcare. Females, who are larger, control territories that encompass those of several males, and quickly produce replacement clutches if any are lost to predators.

INFIDELITY
AND DIVORCE

When females take multiple partners, regardless of their social mating system, sperm competition results. One major way for males to ensure paternity, especially if they have a hard time keeping an eye on their mates, is to swamp their competitors' sperm. Species such as many fairywrens, in which females mate with several males, have evolved larger testes for their body size, enabling males to produce more sperm. Conversely, the male Eurasian bullfinch has tiny testes for his body size. As one might expect, DNA fingerprinting of the chicks in this species has shown that females are almost entirely faithful to their mates. Similarly, wood thrush pairs are highly territorial, with both sexes staying on their territory to guard it. Males are thus seldom cuckolded, and have relatively small testes.

Divorce rates in birds vary from 98 per cent in greater flamingos to a mere 2 per cent in barnacle geese. Just as in humans, some individuals within a population are more prone to divorce than others. Biologists have become very interested in measuring individual personality differences within bird populations. For instance, great tits fall along a bold–shy gradient, depending on how soon individuals are willing to approach a suspicious new object, or how quickly they learn to stay away from something like a trap. Unlike barn owl or jackdaw societies, where infidelity is just not done, individual great tits with bolder personalities are more prone to both infidelity and divorce. It is not clear what causes this variation, but the answer is likely to involve a combination of genetically inherited predispositions shaped by individual experience.

FAITHFUL
BIRD SPECIES

Some species that are largely genetically monogamous include:

- Barnacle goose
- Barn owl
- Jackdaw
- Eurasian bullfinch
- Wood thrush

Opposite, top

Both divorce and infidelity are rare among barnacle goose couples.

Opposite, bottom

Greater flamingos (*Phoenicopterus roseus*) are the most widespread flamingo species and generally bond with their partners for life.

CULTURAL CONFORMITY AND SOCIAL BONDS

While resource distributions directly influence how individual animals are distributed, resulting in different social systems, social bonds also provide feedback on how a society is structured.

Even common garden birds such as great tits have a form of culture, in the sense that flocks establish foraging traditions. Although not as sophisticated as the tool-making styles of crows in different regions of New Caledonia, these still require individuals to learn from, and conform with, the rest of the flock.

Below

Great tits often choose to go hungry rather than feed without their mate.

One way in which new social information can spread rapidly among great tits to create new traditions is for individuals to vary some of their foraging associates. Great tits forgo food rather than be away from their mate. In an experiment using microchips (as for a pet), and programmed bird feeders, biologists prevented mated pairs from eating at the same feeder. Half the feeders closed when a bird implanted with an even-numbered chip landed. The other half were inaccessible to odd numbers. So while one bird ate at a feeder, its partner had to look for food underneath if it wanted to remain with its mate. Results showed that the trailing partners inevitably spent a lot more time feeding with birds they would otherwise seldom encounter if they stuck to the feeders they could access. These great tits clearly put maintaining the pair bond (even if only for a year) above feeding without their mate.

SOCIAL STRESS AND SOCIAL STANDING

Zebra finches are highly social, and mate for life. They use social information to find food, and are able to rely on highly coordinated and cooperative parenting with a long-term

partner – all helpful adaptations for surviving in the harsh Australian desert to which these popular pet birds are native.

Although not cooperative breeders, wild zebra finches live in big groups. Much like the experiments showing that stress compromises song learning in young zebra finches (see page 68), other studies have shown that early stress alters their social behaviour. Specifically, injecting zebra finch nestlings with extra stress hormones made them more discriminatingly gregarious. These birds spent time feeding with more different individuals, including more birds outside their families, and consequently occupied a more central position in the social network. Most other birds in the colony were thus indirectly connected through these central individuals.

Early developmental stress seems to have the same effect on wild zebra finches. In a study from 2019, biologists increased sibling

Above

Zebra finches are highly gregarious birds that feed and breed in colonies.

competition to create stress. They swapped chicks between nests to control for any stress that might be inherited from genetic parents. The biologists thus created low-stress broods of two, and high-stress broods of five to eight chicks.

Through their studies, biologists saw that zebra finches that grew up in larger families behaved like those injected with stress hormones in the laboratory. They spent more time hanging around non-family members and were less fussy about whom they associated with at feeders. These birds thus foraged with more different individuals, and occupied more central social positions.

3

COURTSHIP

Right

Japanese red-crowned cranes perform
a courtship display in the middle
of a Hokkaido winter.

REPRODUCTION OF THE SEXIEST

During the autumn and winter of 2018, a misplaced Mandarin duck, who took up residence in New York City's Central Park, became an international celebrity with the name of 'Hot Duck'. This gaudy male raises the question of why so many male birds sport fancy ornaments, while females are a discreet brown.

Below, left

Drakes like this mandarin duck display their breeding plumage to females in winter, and return to an 'eclipse plumage' in the summer, when the courtship season is over.

Below, right

Long-billed hermit males have longer, sharper bills than hummingbird females, as the result of an arms race between males battling for mates with their bills.

In *On the Origin of Species*, published in 1859, Charles Darwin proposed the theory of sexual selection to explain evolutionary outcomes such as the 'Hot Duck', whose extravagant plumage should act like a beacon to predators. Elaborate feathers also require a lot of the bird's energy to

make and maintain, so should not in theory evolve by natural selection, which requires the survival of the fittest. However, if ornamented individuals attract the most mates, and have the most offspring, that would be reason enough for costly displays to evolve by sexual selection. Female ducks have selected for showy males just as breeders have selected for chickens with elaborate plumes.

The other part of Darwin's sexual selection theory concerns weapons rather than ornaments. Female hummingbirds do not prefer to mate with males that have the most rapier-like bills, but males select on each other through combat – driving the evolution of weaponised bills in hummingbird males, but not in females of the same species.

Anyone who has watched hummingbirds buzzing aggressively round a feeder will know that these tiny, gemlike birds are far from dainty in their approach to highly valued resources. A rich and reliable source of nectar is necessary for females to raise a brood, which they do as single parents. At the same time, it is easy for a single male to guard a resource that attracts females. This leads to a highly polygynous mating system.

Such high breeding stakes for males means stiff competition for the best territories to attract the most females. Male long-billed hermits, a common hummingbird in Costa Rica, are the only birds known to have evolved weapons on their bills through sexual selection in the form of male-male combat. Just as male mammals such as baboons have evolved longer canine teeth (not present in females), the better to slash their opponents with, these tiny hummingbirds use their bills like daggers; males attempt to stab each other in the throat during contests. Adult males have longer, straighter and more pointed bills than females or young males, and the top mandible protrudes beyond the lower one. This makes the bill less efficient for drinking nectar, but deadlier as a weapon.

Both the Victorian public and biologists as late as the 1930s found the notion of male-male combat more palatable than female choice to explain extreme sex differences. Not only was it easier to observe stags or roosters (or humans) competing openly for females, but many biologists also struggled to ascribe an aesthetic sense to non-human animals. However, Darwin studied many examples of female choice in birds. He argued strongly for discriminating females selecting for elaborately decorated males like the 'Hot Duck'. Biologists are still accumulating remarkable examples of how sexual selection has led to impressive courtship displays.

FASHIONS AND SEXY SONS

Fashion can shape evolution just as much as natural selection, but often in more arbitrary ways. Male black grouse display on special dancing grounds, called leks, and females gather to judge. Only the top one or two males win most of the matings, while the majority sire no young at all. Just as with human fashions, hen grouse are susceptible to the preferences of others. Biologists have swayed hen grouse opinion by placing stuffed females in an admiring circle around a male who was previously deemed unattractive, creating a new heartthrob purely by making him appear more popular.

The fashion analogy still begs the question of why females prefer certain ornaments to others. There are three non-mutually exclusive explanations. The first is that male ornaments exploit pre-existing female preferences for something like the colour red. The second is that extravagant ornaments are honest signals of quality because only a really healthy male could afford such an expensive handicap – rather like only being able to afford a Porsche if you are genuinely wealthy. The third, closest to Darwin's original idea, is often known as the 'sexy son hypothesis'. As long as females consistently choose to mate with the flashiest males, and flashiness is heritable, they will have the sexiest sons, and consequently, the most grandchildren. The beauty of this idea is that a purely arbitrary fashion is enough to cause the evolution of elaborate courtship displays, even at a cost to survival.

Female (but not male) zebra finches can be lured into favouring a completely new ornament just by observing the mating success of others. When biologists tied little red feathers to the heads of zebra finches, no one found these artificially ornamented individuals more attractive unless they first witnessed an ornamented bird of the opposite sex already comfortably paired up, next door to a lonely, unadorned individual. Then females would copy each other, and prefer a male sporting a red feather. Males exposed to the same experiment paid no attention.

Barn swallows show that females in the wild do indeed shape their males through mate choice. Different preferences across populations can result in regional differences in male appearance. Experiments on European swallows artificially lengthened or shortened the tails of males, and then measured the number of offspring they sired. Male swallows with the longest tails had the most offspring (partly by attracting more extra-pair females), showing that females preferred the longest

tails, even if they were longer than any a male could grow naturally. Surprisingly, this was not true for the American barn swallow. Instead, American females preferred males with more orange on their bellies, but had little interest in long tails. American male barn swallows are accordingly ruddier but shorter-tailed than their white-bellied, long-tailed European counterparts.

Above and right

Barn swallows are broadly distributed across the globe, including Europe (above) and North America (right), but show regional differences in plumage depending on what local females find attractive in a mate.

SEXUAL SELECTION FOR NEW SPECIES

Birds usually know who to court and mate with by imprinting on their caregivers. But imprinting can go awry; the highly endangered whooping cranes of North America sometimes hang around a group of sandhill cranes, as if believing they are members of the smaller species. This is because there used to be so few adult whooping cranes that conservationists had to use smaller sandhill cranes as surrogate parents.

Many ducks occasionally leave their eggs in another female's nest. When this female happens to be of a different species, the ducklings grow up with confused identities. They later attempt to court the species they were reared with, which partially explains why duck hybrids are so common. A second, darker reason is that forced copulations are common among some duck species (see page 117). Even if males successfully inseminate

Opposite

A pin-tailed whydah in Kenya displays to a
much more sensibly coloured female. Whydahs
and indigobirds belong to the *Vidua* finches
of sub-Saharan Africa.

females of a different species, this is usually
an evolutionary dead end: the hybrids either
fail to develop or to reproduce. However, in a
few cases, innovations in sexually selected
signals among closely related species can
lead to the origin of new species.

Indigobirds form a rich species complex
within the *Vidua* finches, all of which are
obligate brood parasites. This means that
instead of caring for their offspring, all
members of the *Vidua* genus rely on other
species to perform childcare, from incubation
to fledging. However, unlike cuckoos,
indigobirds and other *Vidua* finches are
songbirds that learn their courtship song and
preferences. Indigobirds incorporate part of
their foster father's song into their own
template. Among Cameroon indigobirds,
for instance, males look identical, but sound
like either black-billed firefinches or African
firefinches. This means that when a few
female indigobirds lay their eggs in the nest
of a new host species, all their sons grow up
singing a new type of song; their daughters
prefer this, and so a new lineage is born.
Indigobirds have speciated incredibly rapidly,
probably because of host switches in the past.

Among Darwin's medium ground finches,
similar-sized birds tended to pair up, with
birds learning their mating preferences from
the size of their parents. By chance, two

unusually large individuals immigrated to the
tiny island of Daphne Major – in the middle of
the Galápagos island chain, where they
fostered an entirely new lineage.

Evolutionary biologists Rosemary and Peter
Grant have been studying every bird on this
island since the early 1970s. They dubbed
the male of this pair 'Big Bird', because he
surpassed in body and bill size not only his
own species, but also males of two other
resident species. He sang an entirely different
song, too, which all his descendents inherited,
thereby founding a new species. This new
species has remained reproductively isolated
from all the other medium ground finches ever
since, because no one else is attracted to the
Big Bird song.

It was not until years later that Big Bird's DNA
was sequenced, revealing that he was a
cactus finch from Española Island, more than
100 km (62 miles) from Daphne Major. His
hybrid descendents persisted because their
unique bill shape gave them a competitive
edge over the other resident species. The Big
Bird lineage shows how a new species can
form over just a couple of generations
through the hybridization of two closely
related species, and be maintained largely
on the basis of a unique song.

HANDICAPS AS HONEST SIGNALS

The tails of long-tailed widowbirds are honest signals of quality: at the start of the breeding season, males with longer tails are in better condition than those with naturally short tails.

Biologists have experimented by adjusting the tails on male widowbirds, to see how the lengths altered a male's attractiveness and condition. Males with artificially lengthened tails attracted more females to their territories than males with tails that were the same length that they started with, while males with shortened tails proved the least attractive. However, males with artificially shortened tails lost condition more slowly over the course of the breeding season than males with long tails – maintaining a long tail is clearly expensive as well as sexy.

The trade-off between investing in self-maintenance versus attractiveness also holds for European barn swallows. Males with artificially elongated tails were more attractive to females, but had weaker immune systems than males whose tails were naturally long. Those who cannot genuinely sustain a long tail appear to pay a disproportionately high price for 'faking it', keeping the signal honest.

There is evidence suggesting that female barn swallows prefer males with symmetrical tails. Perhaps asymmetric tails could reflect stress in early development. This shows an interesting parallel with human cultures, in which bilateral symmetry is one of the most attractive criteria for facial beauty. The idea is that stress, particularly early in development, or with age, shows up as a size or shape disparity between features on the left or right.

The chicken order is full of good examples of costly male ornaments. Friends and I have been charged by belligerent male dusky grouse and wild turkeys, and seen their skin ornaments inflate and suffuse with fiery blood in the short time it takes them to reach us to deliver a vicious peck. Cock pheasants with larger wattles and longer tails win better territories and also attract more mates.

Above

Red grouse – Scottish relatives of the North American willow ptarmigan (shown above) – signal dominance and attract females with red cockscombs above their eyes. Male grouse riddled with parasites have smaller combs.

Peacocks attract more females if they have more eyespot feathers on their iconic trains, and chicks sired by peacocks with more eyespots survive for longer. However, this could also be because females invest more in their offspring fathered by a more attractive, presumably higher-quality male. In the related red junglefowl, ancestor to domestic chickens, hens prefer roosters with bigger, redder combs, and lay more eggs after mating with a stud.

We know from a series of experiments involving artificial insemination, and vasectomised but otherwise sexy males, that sexy fathers do indeed bequeath good genes to their sons, independent of maternal investment. Even when a hen did not 'know' the father of her chicks because she had been artificially inseminated, sons of larger combed males grew up to be healthier and have larger combs.

Opposite

Male house sparrows use their black bibs to show status. Males that win more contests with other males grow more black feathers to display their competitive abilities.

SEEING RED AND BEING RED

Many birds prefer red, and the colour is often an honest indicator of quality. Redder male house finches are genuinely healthier; they also attract more females and sire more offspring.

Below

The courtship display of a male magnificent frigatebird involves inflating a red skin pouch at the base of his throat, and waggling the balloon back and forth.

When a male red-backed fairywren transitions from being a nonbreeding helper to the dominant male, his testosterone levels surge, and his bill rapidly reddens to show his increased social status.

One of the reasons that red or orange are such attractive colours could lie in the pigments that produce these colours. These are known as carotenoids (they make carrots orange), and birds use these pigments for all sorts of things. We can roughly separate these functions into a classic trade-off between reproduction and survival. Making bright feathers to attract a mate, or making egg yolks extra orange and nutritious with carotenoids, are both reproductive investments. Reserving carotenoids to boost the immune system is a more direct investment in self-maintenance.

Carotenoids are a handy currency to use for honest signalling. Birds without enough of these limited pigments to spare cannot afford to use them to make their bills or feathers redder. Any bird attempting to advertise their fitness falsely would probably die sooner from illness. Male zebra finches have redder bills when potential mates surround them than when there are no females present – but males injected with an infectious bacterium do not develop a redder bill.

Greater flamingo females get their carotenoids from eating tiny shrimp (this is the reason why zoo flamingos without the carotenoids in their diet look washed out). Unlike most birds, which must invest precious carotenoids while growing their feathers, female greater flamingos literally put theirs into rouge. All birds keep their feathers well-maintained by preening, and many apply a special

conditioner from a preen gland at the base of their tails. Rather than having to invest carotenoids into growing feathers before the courtship season, female flamingos have developed a more flexible technique. They add carotenoids to the preen oil right when they need to display pink feathers, and apply the rouge to their feathers using their cheeks as blushers. Males preen just as much, but lack the extra colour boost. Once the courtship season is over, females stop making rouge as they have to use the carotenoids for other things, such as eggs. This could explain why, in this species, females have evolved a way to be temporarily more colourful than males.

Birds do not just use carotenoids as cosmetics. Unlike mammals, they use the same pigments to see red. In addition to having special molecules sensitive to different colours of light in their retinas, birds have coloured oil droplets in their cone cells. These act as filters to sharpen the distinctions between different colours.

The gene that enables birds to convert the yellowish pigments from their food into red colouring for rouge, skin or feathers is an evolutionary copy of the gene that performs the same pigment conversion for red oil droplets in their eyes. From comparing these genes in the reptilian relatives of birds, we know that the ability to see red evolved before the ability to look red. Only the dinosaur ancestors of modern birds evolved an extra copy of this gene that allows them to look red.

DISPLAYS AND FEATHER EVOLUTION

Birds are the only dinosaur lineage to have survived to the present day. Dinosaur fossils from China have revolutionised our view of feather evolution, as evidence of ornamental feathers suggests that feathers evolved first for courtship displays and were subsequently co-opted for flight.

Birds do not only come in shades of red to yellow from carotenoid pigments, or brown and black from the same melanin pigments that colour human hair. Much of the colour diversity in bird feathers, especially the iridescent blues and blacks, comes from the way in which the nanostructures of feathers reflect light. The process, known as 'thin film interference', is also what makes soap bubbles so colourful.

African starlings are an especially diverse group – not just in terms of the number of species, but also in their plumage colours and social systems. They have evolved all the main innovations in fine feather structure that cause iridescence in the feathers of other lineages such as hummingbirds, sunbirds or birds-of-paradise. Furthermore, the ability to extend their plumage colour palate has driven the origins of new species at faster rates in this group of birds, through sexual selection for attractive new plumage colour combinations.

Iridescence has been under mutual sexual selection for most Old World starlings. This explains why, in species such as the European starling or Asian glossy starling, the sexes are equally shiny. African starlings probably inherited this colourful equality of the sexes, but in the socially monogamous species, such as the violet-backed starling, females have evolved to be much drabber than males. In contrast, superb starlings are

gregarious, cooperative breeders. They form complex, multitiered societies in which females are under strong social selection from their group members to remain almost as colourful as males.

Biologists have also discovered a handful of pigments that are evolutionarily unique to a particular group. The golden-breasted starling, another African starling, has evolved a new way to make yellow. Instead of using carotenoids, like most animals, this species deposits large quantities of vitamin A to colour their breast feathers yellow. Turacos, another exotically coloured African group, have reddish, copper-rich turacin pigments. Parrots, known scientifically as the Psittaciformes, have their own pigments for creating warm colours

called psittacofulvins. Penguins (Sphenisciformes) make spheniscine pigments. These look yellow to human eyes, but also fluoresce in UV light, which birds can see.

Birds have also evolved feathers that act almost like musical instruments and can be played during a courtship display. There is the thrilling, explosive 'boom' caused by the wind rushing through the wings of steeply diving common nighthawks, or the deep vibrations that always remind me of a motor starting up in the woods, produced by ruffed grouse beating their wings while perched on a log. Using high-speed videos, biologists have found out that hummingbirds also sing with their tail feathers. They produce species-specific sounds that range from single chirps to repeated notes.

AVIAN AESTHETICS

Regardless of why they exist, the predilections birds have when choosing a mate can be uncannily reminiscent of what would be called an aesthetic sense in humans. Some of the most dazzling displays integrate visual ornaments with dance and music; they can also involve multiple performers and multiple judges.

Popcorn dancing manakins

Long-tailed manakins perform two main dances. In the 'popcorn dance', two males alternately leap high up above their courtship branch, increasing the tempo, and meowing and buzzing as they rise. In the 'cartwheel dance', males shimmy along a branch with the male in front flipping behind his partner, who then flips backwards in his turn.

The males of many manakin species, such as the long-tailed, lance-tailed and blue-backed manakin, dance and sing in pairs, but there is a clear hierarchy. The alpha male gets all the matings, while the junior partner has a chance to learn from the master. Long-tailed manakin pairs take years to perfect their duets and dances, and females prefer pairs with similar voices. The pair is often joined by even younger, floater males. These birds flit from court to court, but do not perform the actual courtship dance when a female deigns to visit, unless one of the top two males is missing.

Male ducks of many species court females in the winter, using a series of highly ritualised movements that follow a repeatable sequence. This is when female ducks use a combination of showy plumage and superior dancing to choose their mate. It is also when birdwatchers flock to see an exotic duck far from home, misguidedly displaying to females of a different species and getting very little interest.

Wild greater flamingos in the Camargue of southern France are serially monogamous. Both sexes use a complex dance repertoire to select a new mate for each breeding season.

Birds with the most dance moves, and the most choreographic transitions during each display, are the most likely to find a mate. Greater flamingos live for decades, but the most successful dancers and breeders are those in their early twenties; they may display as many as seventeen transitions between eight different postures. By contrast, younger and older birds muster as few as two postures with two transitions, and are much less likely to breed. Not surprisingly, the best dancers tend to pair up.

Dancing flamingos

Dance moves include head flagging, where the birds walk with necks extended to the sky and flip their heads from side to side; a bow with neck outstretched in front and wings partially extended to reveal a flash of red underwing feathers; and rushing first in one direction, then another, in a group as synchronous as a highly trained corps de ballet.

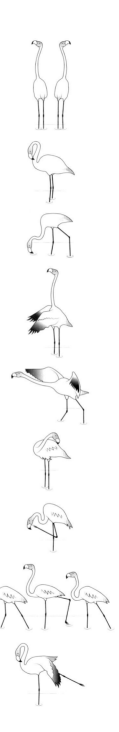

THEY'VE GOT RHYTHM

Members of the parrot family seem to have a strong sense of rhythm. Snowball, a sulfur-crested cockatoo, is a YouTube celebrity: videos show him jamming with gusto and perfect timing to pop music. Although Snowball is a pet, he is not alone in having a fondness for strong rhythms. Wild palm cockatoo males modify sticks and seed pods into musical instruments and drum on hollow branches to attract mates. Biologists studying them have found that individual males have unique styles and named them after famous drummers, such as Ringo Starr. Sometimes, males will accompany the drumming with a few screams, or insert a visual element by blushing bright red on their cheek patches, or erecting their crests.

Christina Zdenek

Above

Male palm cockatoos in the wild drum on hollow branches to attract females, and each male has a distinct style.

MUSICAL BIRDS

Singing is a widespread way for birds to attract mates. Female pied flycatchers and European starlings are attracted to and enter nest boxes from which biologists are blasting recordings of male song. Sarah Earp and Donna Maney wondered if songbird audiences experience something similar to humans as they listen to music. They looked in the brains of white-throated sparrows listening to song recordings and found that for a breeding female, listening to an avian aria triggers the same reward pathways that music stimulates in humans. In other words, breeding female white-throated sparrows may derive the same pleasure from a courtship song as you would when listening to your favourite musician. In contrast, listening to the same songs triggered a

different emotional brain pathway in breeding males. These birds react to songs by a potential competitor in the same way as humans experience music they dislike.

MULTIMODAL DISPLAYS

Courtship displays by birds-of-paradise involve song, dance and the flashing of fancy plumage; females have selected for increasingly complicated multimodal male displays in this famously exotic group. After analysing hours of video and sound recordings, biologists from the Cornell Lab of Ornithology found that males of different species have adapted their displays to the physical properties of their respective theaters. Species displaying high in the forest invest more in complex vocal accompaniments, which can carry easily through the canopy, but have relatively basic choreography – perhaps because they are so precariously perched. In contrast, species displaying in the dank, dense understory use

a wide array of flashy dance moves to attract females, but bother less with vocalising, since the underbrush dampens sound.

There is little wonder that human males of Papua New Guinea are so fond of using the plumes of hundreds of male birds-of-paradise, when the birds themselves have evolved such colourful and diverse ornaments to win the favours of females. Males are much more colourful than females in lekking birds-of-paradise than in species where males perform solo. This is because leks set up a dynamic in which only the sexiest males get to breed. As a result, sexual selection on males is much stronger than in species where female preferences have less power to dictate which genes enter the next generation.

Above

The western parotia from Papua New Guinea maintains up to five display courts. Dance steps include 'puffs', in which a male expands its flank plumes to form a skirt, and 'bill points', where it thrusts its bill forward at a female just before mating.

SELECTING FOR SMART AND SENSITIVE MALES

Bowerbirds are less physically flashy relatives of the birds-of-paradise. Instead, they invest in constructing bowers, artfully adorned with meticulously arranged ornaments to attract mates. These bowers exist purely for courtship; they are not nests, because mated females go into the rainforest to become single parents.

Below

A male satin bowerbird arranges his collection of blue bottle tops adorning the front of his bower in Queensland, Australia.

Satin bowerbirds live for over 20 years, and males take seven years to mature. The aesthetic preferences of female bowerbirds select for smart males. Species with the most complicated bower constructions have the largest brains for their body size. Many incorporate the songs of other species into their courtship displays. Males have to pull off considerable feats of memory and skill to attract the most females, and this can take years to perfect.

Satin bowerbirds also have an innate aversion to red objects, which biologists have used to test for general problem-solving abilities. A male bowerbird knows exactly when anything has been rearranged, misplaced or introduced to his gallery – partly because they often pilfer choice ornaments from neighbours. When experimenters place red objects near a bower, males will do their utmost to remove the offending items, or cover them up. The individuals that solved these experimental puzzles fastest were also those with the highest mating success.

In order to mate, a male bowerbird must eventually lure a visiting female all the way into the centre of his bower. Female bowerbirds are quite skittish; a female will startle easily if a male comes on too strong while she is still inspecting his gallery and assessing his dancing. By using robotic females, biologists could systematically measure how sensitive displaying males were to cues from the females that they were attempting to woo. Through this we know that males that keep displaying ardently to a hesitant, robotic female win fewer matings with real females than the ones that know when to back off, or to tone things down. This suggests that female bowerbirds are selecting for males that are highly attuned to the responses of the females that they are courting.

By using 'fembots' – taxidermied females on wheels implanted with video cameras – Gail Particelli and her colleagues have been getting a female's eye view of displaying male sage grouse in Wyoming and California. They have found that males with the most mating success are those that most flexibly alter the intensity of their energetically demanding booming displays to perform only when there is an audience.

Among long-billed hermits, the hummingbird species in which male competition for territories has selected for spear-like bill tips, males best able to solve spatial memory tests set by biologists are more likely to win a territory. This is true even if they are not endowed with the longest bill tips. These males also tend to sing the most consistent (and attractive) courtship songs.

ALTERNATIVE COURTSHIP STRATEGIES

Not all males are born equally attractive, so rather than sticking to the same methods of wooing females as the top studs, some males employ alternative strategies.

The ruff is a type of shorebird, named for the feathers around the necks of males that resemble Elizabethan collars. Ruffs also lek (gather for sexual display on a courtship ground), but males come in a variety of genetically determined forms, each with its own plumage and courtship strategy. Dominant males with the biggest and best territories have black or chestnut ruffs. On the outskirts of these territories are white-ruffed satellite males, who forcibly attempt to steal copulations with females on their way to visit a top male. Biologists only recently discovered a third male morph, because these males are very rare and look almost exactly like females. These sneaker males invest in disproportionately large testes instead of fancy plumage and strong muscles.

The three male mating morphs coexist because each is best able to persist at a certain frequency in the presence of the others. Dominant males are five times more common than satellite males, but they do tolerate a small proportion of satellites because males in larger groups attract more females. Only 1 per cent of male ruff are sneakers, suggesting that this strategy has just enough success to keep a handful of sneaker genes in the population.

In contrast to the genetically fixed mating strategies male ruff are born with, buff-breasted sandpipers have multiple tricks to acquire mates. These little shorebirds breed in the high Arctic, where conditions are unpredictable and the best lekking locations can change rapidly, both across and within breeding seasons. As a result, males switch flexibly between

Above

The ruff is a species of sandpiper named for its male breeding plumage, which resembles an Elizabethan ruff. Here, two males are displaying competitively.

remaining on one lek all season, moving between leks, and displaying alone or in the company of other males. Males tend to join leks with larger males already displaying, as if to benefit from any reflected glory. Males do not look very different from females, so have to resort to a wing-flapping display to attract attention. Females examine the undersides of males' wings closely, preferring to mate with males with more wing spots; the reason for this is unknown.

By watching about fifty males, each marked with a unique leg band, biologists found that male pectoral sandpipers also lek. However, they only stay on a breeding ground if there are a lot of females available to mate with. Even then, they simply mate and move on, spending less than two days at a single lek. These males are highly nomadic, sampling up to twenty-four leks throughout the species breeding range in Alaska, Russia and Canada, and covering over 13,000 km (8,000 miles) in a month. This is straight after migrating almost the same distance to reach the Arctic Circle.

Sandpiper leks in the Arctic

Pectoral sandpipers gather in leks to find mates, but males can travel all over the Arctic Circle sampling different leks, staying for as little as a day or less if there are few females, to over ten days (in red) at leks with the most females.

DAYS

- 0–1
- 1–2
- 2–5
- 5–10
- >10

MUTUAL SEXUAL SELECTION

Below

Crested auklets of both sexes choose their mates based on a set of criteria ranging from their curly head feathers to a species-specific citrus perfume.

Much of this chapter has focused on extreme adaptations resulting from strongly asymmetric sexual selection. However, many bird species have mutual mate choice, which tends to result in sexes being equally ornamented.

Blue-footed boobies of both sexes choose their partner of years by assessing the brightness of their blue feet, involving a ponderous display of mutual foot lifting. Foot colour could be an honest signal of quality: immune-compromised birds invest fewer yellow carotenoids in their feet, thereby failing to achieve the most attractive shade of turquoise. After only 48 hours of underfeeding, boobies' feet turn a duller shade of blue.

Crested auklets of both sexes attract mates with a variety of signals. These include a curly tuft of feathers protruding from the front of their heads, a fluorescent bill plate, yapping calls reminiscent of a Chihuahua, and a strong, tangerine-scented perfume concentrated around their necks. This scent is so pervasive during the breeding season that biologists near a nesting colony can smell it. The auklet courtship ritual includes a behaviour called the 'ruff sniff', in which individuals press their heads into the necks of potential partners.

This monogamous species starts with a strangely frenzied courtship ritual. It usually involves multiple interested parties in what biologists call a 'mating scrum', where male auklets attempt to suppress rivals' displays. Longer crested males have lower stress levels, stronger immune systems and can produce more perfume. They are also dominant over other males, and more attractive to females.

Great tits with similar baseline stress hormone levels stay bonded for longer and tend to have more offspring. The hormone profiles of longer-term pairs also become more similar over time. Interestingly, stress hormone levels has no effect on levels of extra-pair paternity, so a female might choose to bond officially with a male of similar personality – particularly if she can still mate with other males that have a more competitive personality. Curiously, bolder males – those that in personality tests explore a new environment more readily – form stronger pair bonds with their future mates. This is because they meet their future mate sooner and begin settling down earlier, before the breeding season. Female personality, by contrast, does not seem to alter the timing or strength of these pre-breeding relationships.

Red-tailed hawks renew their pair bond annually in elaborate nuptial flights; these include chases and tumbling in midair while holding on to one another's feet. A coordinated pair is more efficient at hunting. Similarly, golden eagle pairs can hunt cooperatively, with one partner distracting prey while the other grabs it from the other side. Roadrunner parents cooperate to maintain a territory year-round; they build a nest, incubate eggs and feed chicks together.

A FUNDAMENTAL ASYMMETRY

If you are wondering why in most bird species, males are more competitive or showy than females, part of the answer lies in a fundamental asymmetry in parental investment between the sexes.

In biology, females are defined as the sex that produces larger sex cells (eggs), while males are the sex that makes smaller sex cells (sperm or pollen). These early asymmetries in how much the sexes invest in offspring are often amplified, leading to the evolution of vastly different reproductive strategies: females focus their resources on quality over quantity, while males invest in the quantity of reproductive opportunities at the expense of quality. As a result, females are limited by the resources they can invest in a finite number of offspring, while males are limited by the number of matings they can perform.

This difference in reproductive investment strategies explains why females, with fewer, higher quality investments, are typically the choosy sex. Males compete for the relatively scarce reproductive investments of females. However, because all birds have one father and one mother, if a few of the most competitive males win most of the females, the majority of males do not breed. A situation with a few winners and many losers explains why sexual selection is usually stronger on males than females. It also explains why males in species with stronger sexual selection are less inclined to care for offspring. Parental care would take time and energy away from a male's future success as a competitor for mates.

REVERSED SEX ROLES

A handful of birds with showy, competitive females and choosy males are the exceptions that prove the rule of how reproductive investments shape sex roles. Female jacanas and spotted sandpipers both occupy territories that encompass a harem of males who perform all childcare duties. If a nest fails due to predation, the female simply lays a replacement clutch and moves on to the next male. Female spotted sandpipers also have darker spots than males; this

spottiness could be an honest indicator of a female's physical superiority. In these species, females invest less in each offspring than males do. They also remate more easily than the smaller males, which are tied up with rearing offspring.

A CUCKOO REVERSAL

The black coucal of Africa, a species of cuckoo, is the only role-reversed bird known to have highly dependent chicks. As with many other cuckoos, which are famous brood parasites, female black coucals perform no parental duties beyond laying eggs. However, the males

Above

A male African jacana performs all parenting duties, including incubation and keeping an eye on the offspring, while his mate divides her time and eggs between multiple males.

Opposite

A female African jacana bows her head as she courts one of several males on her large territory. Among jacanas, females are the larger, more ardent sex.

of this species do perform parental care. These stay-at-home dads are in sole charge of nesting, incubating eggs and ferrying food to hungry chicks in the nest.

ONE SMALL STEP FOR WOMANKIND

Biologists are finding increasing evidence of competitive females signalling their status with vocal and visual displays, even among socially monogamous birds with conventionally colourful males. The population of dark-eyed juncos that have stopped migrating to capitalise on the clement conditions in southern California, for example, compete for territories all year round. Biologists have observed these female juncos singing lustily, both spontaneously and in response to recorded songs from other females.

It has yet to be shown whether this heightened female rivalry is the unique result of greater competition for resources in birds with a sedentary lifestyle, or if biologists have previously overlooked female singing among migratory populations.

Females are flexibly polyandrous in certain situations. In the case of the dunnocks, monogamy persists not because males are needed, but because within-sex aggression makes it hard for both sexes to aquire extra social mates. In other words, monogamy is the result of a stalemate, because both sexes have the most offspring if they can monopolise multiple mates. Males have most offspring in polygyny, females the most in polyandry.

The outcome of this individual-level conflict is affected by the climate, the habitat on a territory, and the competitive abilities of each bird. Harsher climates often result in excess males, because females are more susceptible to climatic extremes. A male-biased sex ratio leads to a rise in polyandry, because it is harder for males to keep the limited females to themselves. A territory with denser shrubbery also increases polyandry, because a female can more easily evade the guarding alpha male and mate with beta as well. Lastly, older males are more likely to encompass the territory of more than one female, favouring polygyny or polygynandry.

Several cases of polyandry have been documented among woodpeckers. Although most lesser-spotted and three-toed woodpeckers are socially monogamous, biologists have found almost one-tenth of the population living in polyandrous setups. Sometimes, this was because the first male was inadequate. Male woodpeckers usually perform the night shift during incubation, and one three-toed woodpecker found her second mate shortly after the first one failed to appear for incubation duty for a few

nights. Another male was an inexperienced yearling. In both cases, females spent more time caring for their second families, although they would still visit the first.

Similarly, up to 5 per cent of female northern flickers in a population in British Columbia had more than one social mate in a given year. Over the course of 10 years, biologists saw that older, more experienced females were more likely to become polyandrous; their second husbands were often young males who had trouble in finding an exclusive mate. Females reared almost twice as many chicks in polyandry than

monogamy, and spent less time helping to incubate their second clutch. Unlike most socially monogamous birds, female flickers tend to mate openly with more than one male, rather than sneaking off for furtive, extra-pair copulations. Also, unlike other socially monogamous birds, male flickers do not retaliate by caring less for offspring if they suspect their mates are unfaithful – even when she is openly spending time with a second family.

Above

A female red-shafted flicker (the western US variant of the northern flicker) returns to the nest where the chicks await.

POLYANDRY IN EXTREME ENVIRONMENTS

The red phalarope takes role reversals to greater extremes, because it breeds in the Arctic. Food in this harsh environment is abundant for a very short window of opportunity. Since a female phalarope cannot lay more than four eggs in each clutch, the best way for her to capitalise on a brief glut of midges is to lay a quick succession of clutches for a series of husbands. A male's best response is literally to sit tight and tend his nest alone, because the vast quantity of food required creates a situation in which females are limited by how many mates they can find, rather than by the number of offspring they can produce. This state of affairs is self-reinforcing: with males tied up in taking care of chicks, fewer are available for females to mate with. House husbands are in short supply, giving larger, more competitive and showy females an

Below

Two female red phalaropes tussle over a male at an ephemeral pond in the Alaskan Arctic. The male will go on to care for the eggs and chicks of the female who wins him.

advantage. The situation also pays red phalarope females to concentrate all their time and energy on acquiring mates, rather than on childcare.

Among eclectus parrots, females are so much more colourful than males that ornithologists initially classified the sexes as separate species. Being colourful comes at a cost, because the better-camouflaged green males are in the majority. The skewed sex ratio could also explain why these parrots are polyandrous. Females compete with each other for the best tree hollows in which to nest, while males compete with each other to be one of the several mates attending a female. Females with the driest nest holes have the most mates and the most help with childcare. However, because females perform most of the incubation in dim nest cavities, biologists speculate that this allows them to evolve their bright red and blue plumage under sexual selection, whereas males, who spend more time exposed to predators, have evolved to be a well-camouflaged green.

THE CHICKEN AND THE EGG

Many explanations for reversed sex roles, such as polyandry or competitive, larger and more ornamented females, begin with the evolution of paternal care. This begs the question of why fathers sometimes invest in more care in the first place. The answer is complex, but we know that a positive feedback loop ensues. So the sex that invests more in each offspring often also becomes the sex in higher demand. This drives the sex that invests less to vie for access to mates of the more caring sex, increasing the disparity in sex roles between the sexes. The result is that one sex becomes much larger and more competitive; it trys to acquire as many mates as possible. Meanwhile, the other sex is limited by the number of chicks it can successfully raise. It must fend off unwanted attentions and choose only the best partners for a limited number of reproductive investments. If explanations for sex roles seem circular, this is because reproductive strategies do tend to reinforce each other in an escalating 'race' between the sexes.

SOCIAL MONOGAMY AS A STALEMATE

Dunnocks are one of the best examples of sexual conflict over social mating systems. Mating skirmishes start when a female dunnock is lining her nest and last right up to incubation. Alpha males who share a territory with subordinate males have to guard a female earlier and more intensely – junior team members always try to butt in, even when the alphas are in the middle of copulating.

Quarrelling dunnocks

Dunnocks of both sexes squabble over sole access to their mates, and the benefits that come from exclusive help with childcare.

A female dunnock identifies males by song; if she hears the beta male serenading her, she will do her best to meet him. She actively flirts with subordinate males, hopping up to present her rear end with tail raised and wings coyly a-flutter. Guarding is costly for males; they cannot take their eyes off a female to feed without risking her consorting with subordinate males. It is also costly for females, as a mate dogging every step cramps her ability to forage.

A female dunnock may lead the alpha male a merry dance through vegetation, then hide quietly in the shrubbery with a subordinate male while alpha searches for them. This is so common that when a female has evaded both her consorts, the alpha male follows subordinate males to find her again. Only in a situation where both sexes have multiple partners, would an alpha male cease constantly guarding a single female and divide his attentions between several, allowing subordinates and neighbours plenty of opportunities.

Why do dunnock females seek to mate with more than one male? When females have more than one mate, frustrated subordinate males may sabotage the nest she shares with the alpha male. In contrast, if a female mates with all available males, aggression ceases when incubation begins. Furthermore, a dunnock male helps to feed chicks if he has mated with their mother. As a result, females that mate with more than one male have more childcare help; they can thus afford to lay a larger clutch and raise more offspring. However, if males have multiple mates, the females have to share paternal help, and raise fewer offspring than females that can monopolise a male.

Male dunnocks are not the only ones squabbling over exclusive access to mates. Female dunnocks fight each other, especially when sharing males. Females in polygynous relationships often abandon eggs because of persecution by the group's other female. Males try to break up fights and chase both females back to their territories; males remain in between, feeding female neighbours, and maintaining an uneasy truce. When an artificial feeder was shared by two female dunnocks, the male who overlapped both territories would side with whichever female was laying eggs, allowing her more time at the feeder.

Males do not benefit from cooperating: they only live a few years, so only invest in caring for chicks if they have some chance of siring them. Males mated to one female will increase care in proportion to their mating share (and probability of paternity), which they seem to estimate by keeping a close watch on their mate and on her other male partners. A female gets the most care for her chicks if she manages to divvy up mating equally between two males, as then three adults care for one brood.

SPERM COMPETITION AND CRYPTIC FEMALE CHOICE

Although 90 per cent of birds form socially monogamous partnerships, the vast majority also engage in what biologists call 'extra-pair copulations', only some of which result in offspring. The superb fairywrens of Australia win the record for infidelity, with an average of two-thirds of chicks in a nest being sired by a male other than the female's social mate. As a result, sexual selection and conflict also take place within the reproductive tract.

Fairywren species that experience more sperm competition have evolved a longer 'cloacal tip' – a muscular protuberance that extends beyond the hole called a cloaca through which most birds mate. In response to potential female infidelity, male dunnocks peck hard at a female's cloaca when she is soliciting a mating. This stimulates her to extrude globs of sperm from any previous mating, which the male inspects before proceeding to mate himself. In the closely related Alpine accentor, females have

overlapping territories; they actively court males by singing and displaying bright red cloacas, leading to very strong sperm competition between males. A single female accentor has been observed copulating over a thousand times before laying a clutch of eggs. Unlike dunnocks, males do not force females to eject the sperm from previous copulations. They have also evolved larger testes and copulate more frequently than dunnocks.

Males of the red-billed buffalo weavers of Africa have evolved a non-erectile phallus in the form of a stout muscular nubbin. Buffalo weavers nest in colonies, and males breed cooperatively, often building nests together and sharing mates, territory defence and chick-feeding duties. Males can be the genetic fathers of chicks both within and outside the cooperative group, so sperm competition is intense.

Mysteriously, their phallus contains no sperm and remains completely dry after copulations. Ornithologists puzzled by this phenomenon observed male buffalo weavers engaging in another highly atypical bird behaviour. A single male would mount a female multiple times, with each copulation lasting for minutes (instead of the few seconds that most songbirds bother with). At the end of this lengthy mating process, males appeared to experience an orgasm, in which their entire body shook, their wingbeats slowed to a quiver, and their feet clenched in spasm, with the effect of pulling the female tight against the male at the point of ejaculation. This bizarre behaviour constitutes the only known avian orgasm as a necessary precursor to ejaculation. It seems to take males about half an hour of repeated mounting and copulation to achieve the requisite stimulation.

Researchers did go so far as to attempt to stimulate males with their fingers, but although they got close, the buffalo weavers never quite achieved orgasm and ejaculation in human hands. The best explanation these researchers could give for the evolution of such a convoluted copulatory ritual was that it might keep females occupied for longer, thereby functioning as a rather energetically demanding version of mate guarding. The paper makes no mention of the females' responses to any of this energetic and lengthy mating, so that side of the equation remains a mystery in this species.

Opposite

A female dunnock solicits copulation with her tail raised and wings aquiver, while the male eyes her suspiciously. Before mating, he will peck at her cloaca to ensure that she extrudes any sperm from a previous male.

ALL'S LESS FAIR WHEN LOVE IS WAR

This chapter starts and ends with ducks. Sexual conflict can give rise to some rather ugly outcomes when males and females are stuck in a battle of the sexes. In dunnocks, skirmishes can take place during the lifetimes of individual birds, playing different strategies every breeding season. In ducks, they play out over generations, culminating in the evolution of cumbersome genitalia instead of the simple cloacas that most birds have, designed for copulation that lasts a split second.

My first encounter with the darker side of bird courtship occurred while picnicking on the banks of the River Avon, in England. The bucolic scene was shattered by a whirling mass of mallard ducks that moved rapidly downstream towards me. Even more startling was the fact that any males sitting peacefully near me promptly made a beeline for the melee. At the heart of this increasing mass of males was a female mallard, struggling futilely to escape. Every time she raised her head for air, the males on or near her would stab at her head and grip her neck, and attempts to dive would cause the scrum to pursue her underwater. By the time I lost sight of this mob as they drifted down the river, there were at least fifty males in hot pursuit.

I later found out that some ducks are infamous for what biologists call 'forced copulations' late in the breeding season, when there are a lot more males than females around. This is probably because many females have been picked off by predators while nesting. The last thing the survivors want to do is raise yet another brood, especially since they have already reproduced with a genetically superior male of their choice. However, the majority of males are now at a loose end, and lose very

little from attempting to score one last attempt at breeding late in the season, especially since the females do all the work of incubation.

The physical evidence left behind from generations of these forced copulations are some positively baroque genitalia. Species like ruddy ducks and mallards (ancestors of the domestic Peking duck), with the most sexual conflict, have the longest phalluses. Females of the species have correspondingly longer vaginas. These are filled with dead ends or are corkscrewed, but in the opposite direction from the corresponding phallus. In 2009, Patricia Brennan ingeniously demonstrated the efficacy of these evolutionary chastity belts by making test tubes coiled in different directions and then filming insemination attempts at high speed. For this, she found drakes from a foie gras factory that were used to being sperm donors.

These duck species are in a long-standing arms race between males evolving longer and longer phalluses, and females who counter-evolve defences to retain the freedom to choose a mate. In contrast, buffleheads, wood ducks and Mandarin ducks (like 'Hot Duck') form stable pair bonds that can last years, and both sexes have much more modest genitalia. Ultimately, all the drama of courtship revolves around how to maximise one's genetic investments in offspring.

Opposite

Four mallard drakes congregate on a female in their attempts to mate with her. These forced copulations can result in the female drowning.

FAMILY LIFE

Right

This blue-footed booby chick may
well end up killing its as-yet-unhatched
younger sibling.

COURTSHIP OR CARE?

In evolutionary terms, mates are like joint genetic investors in the next generation. Sexual conflict between reproductive partners can continue over who gets left holding the babies.While lekking male sage grouse or birds-of-paradise invest all their reproductive efforts into courtship, for example, most birds – both male and female – must also save enough energy to raise their offspring. The inevitable trade-off between courtship and care can lead to different outcomes across individual lifetimes, and across the evolutionary history of entire lineages.

Most birds form socially monogamous pair bonds because it takes two adults to rear a brood successfully. Unfortunately, males do not always help with childcare. Collared flycatcher males, for example, tend to set up shop at nest boxes more than 60 metres (200 feet) apart, possibly to minimise aggression between females. A female may be deceived into mating with an already mated male. Once the male has mated with this second female, he deserts her, returning to help raise his first family. She raises 40 per cent fewer chicks than if she had had help.

However, penduline tits and Kentish plovers are examples of species where single parents still manage to pick up the slack. Between 30 and 40 per cent of penduline tit clutches fail because both parents found better options, absconding before incubation has even commenced. Among Kentish plover couples, females often leave first in populations where they are in the minority. This is because a male-biased sex ratio means it is much easier for females to remate than males. In addition, young plovers are highly precocial. They are covered in down and able to run within hours of hatching, so they are more likely to survive with a single parent.

At the other extreme, Laysan albatrosses cannot fledge a chick without two adults, taking turns to incubate a single egg and then to feed the chick. Widowed Laysan albatrosses form strong pair bonds with each other that can last years. The two females function much like a heterosexual couple when raising the chick. However, one of the widows has to let her egg go unhatched, because a pair has the resources to raise only one chick.

SEXUAL EQUALITY

In a comparison across 650 bird species from over 100 families, biologists have found that parental care by the two sexes is most equal when sexual selection (sexual dimorphism and extra-pair paternity) is weakest, and adult sex ratios are the least skewed. In other words, species with the most egalitarian sex roles are also those in which neither sex finds it much easier to find additional social mates or have extra-pair copulations. This could create a positive feedback loop whereby since neither sex is in excess, there is no asymmetry in mating success: both sexes compete equally for a reproductive partner. This causes the two sexes to look and behave alike, giving both an equal chance of surviving, which leads back to even adult sex ratios.

Opposite

A Eurasian penduline tit returns to its nest. Mutual desertion in this species can cause up to 40 per cent of nests to fail.

FAMILY BUSINESS

Raising a family may involve short-term negotiations over who cares, how much to care, and who to care for. Negotiations can take the form of ritualised songs and dances between couples, but both sexes can also invest differently, depending on the perceived quality of their genetic partner.

Blue-footed booby couples dance

Blue-footed booby couples tend to stay together for years, and go through an elaborate courtship ritual every breeding season, which revolves around the mutual admiration of bright blue feet.

Birds such as blue-footed boobies or albatrosses need a couple to raise offspring successfully. They often remain bonded for years, reaffirming their pair bonds with ritualised duets and pas de deux. Blue-footed boobies engage in mutual foot displays. Western grebe couples look most improbable when they hoist their ungainly bodies clear out of the water and streak across the surface, their long necks held exactly parallel to each other. Geese of various species wave their necks and cackle in unison as part of a 'triumph ceremony', after a couple has successfully ousted an intruder.

These rituals may help to cement pair bonds – a good idea if long-term couples raise more offspring than serial monogamists. Southern pied babbler pairs that have been together for longer have the most stable groups of helpers at the nest, and raise the most offspring. Similarly, long-term blue-footed booby couples raise 35 per cent more offspring than newlyweds do. Couples with the most egalitarian parenting are less likely to divorce.

Left

Many grebes, including this pair of western grebes, perform courtship pas de deux, in which partners synchronise their movements with balletic precision.

In songbird species that live in harsh environments where single parenting is simply not an option, both partners benefit from coordinated care. Zebra finches, native to the Australian deserts, duet to coordinate when to swap incubation duties. When biologists temporarily hold up a male during his food break, his return prompts the female to initiate a much more rapid and truncated duet – signalling that he has taken longer than usual and needs to make up the time. After taking her break, the female spends a shorter time incubating on her next shift. The male zebra finch seems to know this, and returns earlier than usual to relieve her.

Another reason why pairs could be synergistically successful at parenting is that males and females do not always use the same signals of need from hungry chicks. It could also be easier to rely on one's partner than on the chicks themselves. Great tit males only bring more food if they see more open mouths, for example, whereas females respond to both visual stimuli and vocal cries. However, if biologists enhance the amount of begging a female hears with playbacks, both parents increase their food delivery rates. Presumably, a male great tit increases his own food delivery in response to his mate's behaviour, which makes sense if females are better informed about the level of chick demand.

Both parents could benefit through rearing more chicks if they can share information about what chicks need. Unlike the great tits, superb fairywren males responded to experimental playback calls that doubled the volume of begging by bringing in more food. Females, by contrast, did not increase their food deliveries, perhaps because they were waiting for more cues to confirm the need for more food.

HEIRS AND SPARES

Parents can even set up a situation where sibling rivalry does the hard work of culling the weakest for them during hard times. Birds of prey, and seabirds such as boobies, have more offspring than they are likely to raise successfully, and begin incubation as soon as the first egg is laid. This staggers the age of the chicks, from the first to hatch to the runt of the brood. This last is often called an insurance chick because it only survives if one of the earlier-laid eggs fails to hatch.

Nazca boobies take this to the extreme with obligate siblicide: one of their twins invariably kills the other. Blue-footed boobies are a little less dramatic, and if there is enough food, both chicks stand a chance of surviving. However, a mother can stack the odds against the second chick by laying a smaller egg if she sees that her mate's feet have turned an unattractive dull shade of blue – either because he has genuinely lost condition, or because biologists have painted his feet as part of an experiment. Interestingly, the chicks are remarkably robust to being bullied from an early age. As long as they survive, blue-footed boobies that were bullied as chicks are just as successful at reproducing as their larger siblings.

Squabbling and siblicide are common among eastern screech owl chicks, even when prey is plentiful. Similarly, the endangered Madagascar fish eagle was only ever observed to raise a single chick in the wild, although the pair usually started with two. In 1977, an ornithologist reported watching a lesser spotted eagle chick so cowed by its older sibling's bullying that their mother had to repeatedly coax it to accept food.

Barn owls also have more offspring than they are likely to rear successfully. Here, females strategically vary the timing of egg laying with the phase of the moon, depending on the colour of their mates. During the breeding season, males are the main providers for the family, so their hunting success makes a big difference to the number of offspring a pair can fledge. In 2019, biologists found that on moonlit nights, white males bring in more prey than their reddish counterparts.

Above

Barn owls begin incubation with the first egg laid, so the first chick to hatch has a substantial head start over younger chicks, setting up a highly uneven playing field among squabbling siblings.

This is not because they put in more hunting effort, but because their main prey, voles, respond to owls by freezing like deer in the headlights. They freeze for longer (making them easier to catch) when confronted by the glowing white feathers of white barn owls amplified by moonlight. On moonlit nights, when red fathers are at a disadvantage in the hunting arena, their chicks are smaller and put on less weight than the young of white males. The youngest chicks are the most vulnerable to this inequality. So depending on when they hatch, chicks born to red fathers can end up with less food and more sibling competition than those with white fathers. Remarkably, females with red mates start laying their eggs at lower levels of moonlight, so as to minimise the time their chicks have to grow up with a less effective hunter as a father.

Below

Nazca boobies are obligately siblicidal, meaning that one twin always kills the other. This older chick is hogging the parental shade and pushing its younger sibling into the sun.

SIGNALLING WITH NESTS

Commitment is an unenforceable contract, so having a male provide a nest as part of his courtship enables females to guarantee some paternal contribution before mating. In many hole nesters, from Atlantic puffins to red-breasted nuthatches, males begin most of the excavation. The female then finishes off if she approves of his work.

Black wheatear males weigh less than a golf ball, but still carry up to fifty times their body weight of stones to their nests. Males that build heavier stone piles have more offspring. When biologists helped males by adding stones to their piles, females responded by laying eggs earlier. Biologists have found no clear evidence of any structural or protective function, suggesting that the stones' primary purpose is to signal investment quality to a potential partner.

Other nest signals include feathers and plants. Blue tit males carry feathers to a nest they have created, but leave the female to line the interior. When biologists add extra feathers to a nest, attempting to simulate the appearance of another male's visit, the resident male responds to the possible risk of cuckoldry by reducing their own childcare efforts.

Conversely, when female blue tits bring lots of aromatic herbs into the nest, their mates take more risks to attend to the chicks, and the pair raise more fledglings. The herbs may serve to deter parasites, but this has yet to be clearly demonstrated.

HOME MAINTENANCE

Birds keep their nests clean in many ways, so as to deter both predators and parasites. Songbird caregivers collect faecal sacs from their chicks' rear ends and deposit them far away from the nest. Hornbill, hoopoe and hawk chicks aim their projectile droppings out of the nest altogether. Some birds even use pesticides or antimicrobial substances in and around their nests. House finch and house sparrow nests littered with cigarette butts contain fewer lice.

SOFT FURNISHINGS

Birds often use soft insulating materials to line their nests. Common eiders – producers of the original eiderdown – nest in and near the Arctic, and use down plucked from their own breasts to line their nests. In the seventh century, an early example of conservation law was enacted by St Cuthbert, to protect an eider colony in the Farne Islands, off the northeast coast of Britain, from theft of their down. Over a thousand eiders still breed in the Farne Islands colony, bearing the local name of

'Cuddy duck' in their saint's honour. Humans still harvest the down to fill expensive quilts and coats, but it is now harvested sustainably, and only after the chicks have left a nest.

Above

Male masked weavers construct over twenty nests in a breeding season. They advertise for mates by hanging upside down from their creations while singing and fluttering their wings. Skilful weaving takes practice, with only nests that pass muster being accepted by females. Rejected nests are often dismantled.

EGG EVOLUTION

Nests are essentially extended wombs in which to keep eggs safe, and in the case of birds with dependent young, the chicks as well. Rather than undergoing a lengthy gestation to give birth to live young, female birds deposit their reproductive investments at an earlier developmental stage: as eggs. This strategy tends to equalise the sexes, because it is harder for males to abscond if they have to remain until eggs are laid to ensure paternity. At this point, their joint investment is sitting outside their partner's body, giving the female an almost equal chance of deserting first.

Male ostrich incubating eggs

Male ostriches share incubation duties with a dominant female, who lays more than half the eggs in a communal nest, which also contains the eggs of other females and males.

Early birdlike dinosaurs were already laying their eggs in nests. Oviraptor dinosaur eggs from China, dating from the Late Cretaceous era, have the same pigment structures as modern birds' eggs fossilised in their shells. Their nests resemble those of ratites (ostriches and their relatives), which contain the eggs from multiple females, all incubated by a single male. Palaeontologists suggest that these blue-green eggs may have resembled emu eggs, matching the background colour of the nests for camouflage. They also speculate that dinosaur females could have been under selection to lay blue eggs to attract male investment in incubation, long before the evolution of modern birds. In 2018, Jasmina Weimann and her colleagues fired lasers at the eggs of another fourteen dinosaur species from lineages that gave rise to modern birds. They discovered in them the same pigments that make chicken eggs blue or brown.

Many of the same hormones underlie parenting across many species and both sexes. Biologists have found that prolactin – the hormone that controls milk production and a lot of the strong maternal instinct in mammals – is responsible for driving the paternal instinct that male penguins have to incubate eggs. When biologists block prolactin in male Adélie penguins, the eggs are incubated for less time at lower temperatures, and the pair raises fewer chicks.

In contrast to the constant care and brooding by parent penguins, fork-tailed storm petrel embryos are often left alone on small islands in the North Pacific. They are so hardy that they can withstand days of neglect, however, even at average temperatures of 10°C (50°F). Storm petrels fly far out into the Bering Sea to feed, often leaving their nest for five days at a stretch. The embryos' solution is to delay hatching by experiencing a form of embryonic torpor within the cold egg. Most eggs are completely neglected by both parents for a total of eleven days, out of an average incubation period of forty-six. But these times can be significantly longer, with the current record holder spending thirty-one days unincubated and taking seventy-one days to hatch.

NATURE'S EGGSTRAVAGANZA

The dinosaur lineages closely related to modern birds typically buried their eggs. Some of the earliest evidence that dinosaur parents cared for their offspring comes from Montana, where groups of fossilised eggs and chicks from the Late Cretaceous period were found clustered in a nest. Palaeontologists believe that the eggs were incubated by the heat from rotting vegetation in the nests. The oviraptor dinosaurs, represented by *Heyuannia huangi* from China, evolved open nests, much like the ratites – a group that includes all the large, long-legged flightless birds such as ostriches and emus.

In most mammals and birds, males are larger and showier, more aggressive, and more ardent than females, who tend instead to be coy, choosy and caring. However, this state of affairs would be far from the norm for most fish, while among birds of prey, females are the larger sex. Similarly, one could be forgiven for assuming that the bright red and blue eclectus parrots are males, and the plain green individuals females. But this is not the case: in this species, females are more colourful and competitive than males. Males' contributions to care probably occurred early in bird evolution.

Ratites and another ground-nesting group, the tinamous, branched off earliest in the family tree of modern birds (see Palaeognathae opposite). They incubate their eggs in large, communal nests, with a single ostrich nest containing about twenty eggs laid by different females. The eggs are incubated by a male and a dominant female. However, the latter often rolls the eggs of other females out of the nest, to make incubation more manageable. Greater rhea females also move between the nests of multiple males, who are in charge of incubating as many as eighty eggs. Although the eggs of an individual nest may be laid as much as two weeks apart, they all hatch within two days of each other – possibly because chicks coordinate the process by calling from inside the egg. Tinamou males also incubate all the eggs from multiple females that lay in multiple nests. These males sit so tightly on the eggs that researchers can pluck a feather from their bodies before they move.

Above

This elegant crested tinamou and its relatives are members of the Palaeognathae, which includes ratites such as ostriches, emus and rheas.

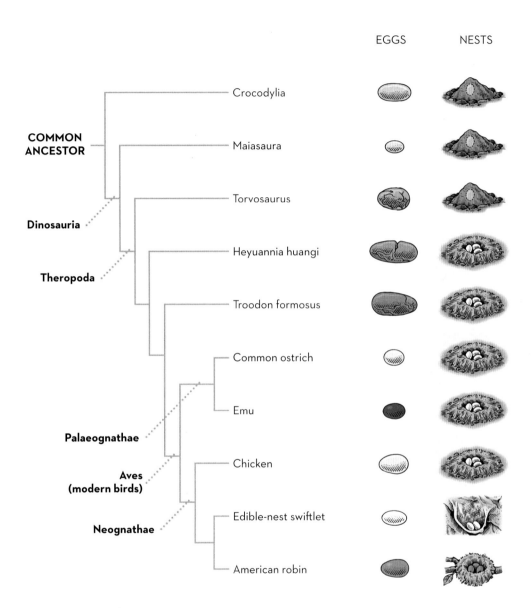

EGGS NESTS

COMMON ANCESTOR

Crocodylia

Maiasaura

Dinosauria

Torvosaurus

Heyuannia huangi

Theropoda

Troodon formosus

Common ostrich

Emu

Palaeognathae

Aves (modern birds)

Chicken

Neognathae

Edible-nest swiftlet

American robin

Evolutionary tree of eggs and nests

Blue eggs evolved independently in the
oviraptors, emus, tinamous and songbirds
(represented at the bottom by an American
robin's egg).

COMPETITION AND MANIPULATION

Obtaining the optimal level of care is important: offspring that receive the most care are most likely to survive to reproduce themselves. This means that nestlings are under strong selection to compete directly with each other, and indirectly by extracting as much as possible for themselves – often at the expense of their caregiver's future reproduction.

An exhausted parent or nanny is likely to take longer to reproduce than an adult that is not constantly coping with young ones.

Adélie, chinstrap and gentoo penguins keep their chicks in crèches. When they return with food, family members recognise each other by their voices, but instead of surrendering their fish immediately, parents turn around and run away, hotly pursued by their demanding twins. Parents pause briefly to feed a chick who has managed to catch up, but take off again as soon as the other chick approaches. This madcap feeding process is inefficient, taking a parent 10 to 15 minutes to feed both chicks, because it has to divide about 1 kg (2 lb) of fish into fifteen to twenty regurgitated feeds, interspersed with short sprints. Any attempt to feed two chicks at once would result in wasted food, because no one stops to retrieve food dropped in the scrum. One function of these chases could be brood dividing, to minimise competition between chicks.

Among bird species with highly dependent chicks, those that experience more cuckoldry also have louder, more selfish begging calls. This could be because the chicks sharing a nest are more likely to have different fathers, so have less genetic investment in the long-term welfare of their nest mates.

If chicks could identify relatives and only make more selfish demands for parental attention when surrounded by unrelated chicks, that would maximise the chances of their genes being passed on. Louder, more selfish cries essentially manipulate adults into investing in chicks that might not even carry their genes, at the expense of their own current and future offspring.

To see if zebra finch and blue tit chicks reduce their selfish clamoring in relatives' presence, biologists have conducted ingenious experiments involving the little-appreciated avian sense of smell. First, they placed broods of nestlings in socks, to capture their scent. Then they attached the scented socks to squeezy bottles, rather like those you pump mustard or ketchup from in a café, and puffed nestling-infused air into a chamber containing a solitary, hungry chick. They chose the smallest nestlings for this, as least likely to be the product of an adulterous liaison. The chicks were exposed to puffs of air 30, 90 and 150 minutes after food deprivation to see how much they would beg while smelling a familiar or foreign brood. Chicks bathed in the scent of the brood they were raised with did indeed beg less than when surrounded by a foreign scent. We do not know yet if this was because chicks learn the scent of a familiar brood from a very young age, or if they are genetically programmed to recognise their relatives' smell.

Fast food

Adélie penguin parents have their chicks engage in a race for each regurgitated food offering by turning around and dashing off upon their return to the nest.

CHICK APPEAL

As noted above, the first evidence that some birdlike dinosaurs cared for their young came from fossil nests found just east of the Rocky Mountains, in Montana. These dinosaurs are named *Maiasaura*, meaning 'good mother lizard', because the nestlings were clearly too large to have just hatched, and were probably being fed by their parents. In addition, their heads had shortened snouts and larger eyes, much like puppies or kittens, suggesting that these baby dinosaurs may have evolved to look endearing to attract more parental care.

Avian chicks can also signal their health by displaying redder mouths, which makes them a more attractive investment to parents. As one would predict, if redder gapes are costly signals that stimulate parents to deliver more food, only helpless, altricial chicks in well-lit nests invest in them; nestlings in dark nests have paler gapes. The carotenoids that chicks use to make their open mouths more appealing are the same precious pigments used as honest, sexual signals of quality by adult birds. When biologists tried supplementing New Zealand hihi chicks with extra carotenoids, the chicks had redder gapes and received more food. However, when the parents were also given carotenoid supplements, they did not increase the amount they fed to their ruddier-mouthed chicks, but instead invested in a second brood.

Young American coot chicks start off with bright-orange plumes and red skin on their heads. They use these features to extract food and attention from their drab, black parents. We know the ornaments are costly because chicks hide their bright heads at the first sign of danger, yet they seem to have evolved under a regime of parental favouritism for the most ornamented chicks. This could be because the brightest chicks are honestly signalling genetic quality or need (the youngest tend to be the most ornamented), or it could be the chicks' way of exploiting preexisting parental sensory biases to receive more attention.

American coot chicks are under especially strong selection to elicit care because females also dump their eggs in each others' nests. There is less of a genetic cost to being a selfishly successful beggar at the expense of unrelated nest mates. In addition, adult coots are under selection to resist the allure of lavishing care on parasitic offspring at the expense of their own genetic descendants. This sets up an evolutionary race between adult coots, which evolve to become ever harder to please, and chicks, who evolve to be ever more appealing.

Above

The internal bill pattern decorating this young, parasitic, purple indigobird's gape mimics that of its host species, the Jameson's firefinch.

VOCAL TUNING

Many bird vocalisations are not only genetically influenced, but also require the appropriate learning environment to develop. This allows some birds to use sound flexibly and strategically.

Communication systems can be hijacked for nefarious ends – such as when cuckoo chicks evolve to mimic the sight or sound of an extra-needy brood, to obtain the maximum care from unrelated host parents. Superb fairywrens have evolved a counter defence to detect imposters in the nest. Although Horsfield's bronze cuckoo chicks closely resemble young

superb fairywrens, even evolving tufts of down that resemble host chicks, adult fairywrens often reject them. This is because female superb fairywrens use a special incubation call: key password syllables are learned both by other caregivers in the group and by developing chicks listening through the eggshell. Cuckoo eggs, laid a few days later, miss learning the crucial password, so fail to incorporate it into begging calls. This allows adult fairywrens to identify foundling cuckoos and to feed only fairywren chicks.

Zebra finch females also call to their chicks through the shell. In this case, they use the incubation call only in years when temperatures rise above 25.5°C (78°F). This unique call slows the embryos' development, increasing their chances of fledging. The early preparation for hotter conditions appears to persist, as females that heard the special call through the shell have more offspring, and prefer to breed in warmer temperatures.

In addition to preparing their offspring for climatic conditions, songbirds such as zebra finches serve as vocal models for young birds. Young zebra finches raised in sound isolation chambers develop a highly jumbled song; it bears little resemblance to the mature song of a finch that grew up among adult models. Yet in a few generations of cultural evolution – and much to biologists' surprise – the descendents of these isolated zebra finches slowly converge on a more sophisticated song structure, similar to that of wild zebra finches. If the first generation of birds deprived of vocal models is allowed to sing for their offspring, the second generation starts to produce a song a little more like that of their wild ancestors. The third generation's song is even more sophisticated, and so on. In only a few generations, a population of zebra finches develops a song that resembles the original much more than the chaotic song of the socially isolated generation.

In another experiment on male Bengalese finches (the females do not sing), biologists showed that genetic differences can influence a song's tempo. Young males learn most accurately to produce a song if it matches their father's tempo. To ensure other aspects of the rearing environment did not influence song learning, biologists swapped eggs, so that young finches were not raised by their biological parents. Later, these youngsters only heard songs generated by a computer. They reproduced tutor songs most accurately if they were played at a tempo closest to that of their genetic fathers. This suggests that while the content of Bengalese finch songs can, and does, evolve and vary culturally, the song's speed is under some genetic control, and influences an individual's ability to learn.

STAYING HOME

Many people in developed countries can identify with the parental problems that face birds whose young delay dispersal. To what extent does letting one's adult offspring remain in the family nest help or hinder their future success?

Species or populations with adult offspring that remain at home tend to live in stable environments, where annual adult mortality is low – resulting in a very low turnover of territories. They also tend to take longer to become competently independent, even after reaching sexual maturity, so there are benefits to remaining in the parental home. Rather than simply delaying dispersal because the housing market is saturated, young can gain materially from parental nepotism by inheriting a territory.

MAKING THE BEST OF A BAD JOB

Western bluebird males assist their male relatives if they are unable to breed, either because their nest fails, or they cannot acquire a mate or a breeding territory. Contributions to rearing siblings do not compensate for the loss of independent breeding, but they are better in evolutionary terms than passing down no genes at all. In western Montana, nest holes are the limiting resource. Putting up nest boxes has resulted in more independent breeders, and fewer sons returning from migration to help their fathers. In the California mountains, western bluebirds defend year-round territories that include mistletoe plants, upon which they depend in the winter. If mistletoe berries are in short supply, or are removed by biologists, fathers chase their sons away. However, like Florida scrub-jays, parents with larger territories can often benefit their sons by allowing them to occupy a portion of the family estate.

NEPOTISM

Many human parents have help with childcare. Approximately one-tenth of bird species are cooperative breeders, in which more than two adults tend a single nest.

Opposite

A male western bluebird located in Troy, Montana about to ferry food back to the nest.

An overwhelming majority of cooperative breeders help and accept help from genetic relatives. This is particularly convenient for white-fronted bee-eaters, sociable weavers and long-tailed tits, all of which nest close together, and for Galápagos mockingbirds, western bluebirds and bell miners, which have fluid territorial boundaries. In all these species, siblings of one sex (usually males) settle close to their parents. The upshot of this limited dispersal is a kin neighbourhood: just by moving next door, one is likely to be helping a relative.

Carrion crow parents are more indulgent to their adult offspring than to immigrants. For instance, dominant males keep immigrant male cobreeders away, enabling their adult sons and daughters to eat before their superiors in the group hierarchy. If dominant males have bold personalities, they spur their shy children to explore and try new food items. Among species such as southern pied babblers and superb fairywrens, sons stand to inherit a position as a breeder when the dominant male dies. They are more likely to stay as helpers than their sisters.

Conversely, birds that breed with nonrelatives experience more conflict within groups. Nest robbery, egg desertion and infanticide are all much more common among species such as Mexican jays and anis, all of which nest with nonrelatives.

FLEXIBLE FAMILY LIFE

Carrion crows stay and help in wealthy, saturated neighbourhoods, but not in poorer ones. Populations of carrion crows breed cooperatively in northern Spain and urban Switzerland, for example, but not in most of their European range. This is a response to the environment, as shown by biologists.

Cooperative carrion crows

Carrion crows are flexible, cooperative breeders, so when real estate is saturated, adult offspring remain to help the breeding pair, whereas nuclear families form when there is enough space for everyone to have a territory of their own.

In experiments, they transplanted half the eggs from pairs of carrion crows breeding in rural Switzerland to the nests of cooperatively breeding pairs in northern Spain. They found that most of the chicks that grew up in Spain stayed to help their foster parents, while none of their siblings who remained in Switzerland delayed dispersal. When biologists raised the property value of territories with supplementary cans of dog food (a favourite among crows), they found that adult offspring delayed dispersal longer than at unsupplemented sites.

Cooperatively breeding carrion crows form groups of up to nine adults. All adults assist with defending a year-round territory, nest building, chick feeding and removing faecal sacs from chicks in the nest. Only one female breeds, but there is DNA evidence that in addition to her social mate, the dominant male, she also mates with immigrant males, who are often his cousins. Adult offspring remain on the family estate and help out for up to four years, but do not breed.

Carrion crow mothers with more help lay smaller eggs, and focus more on self-maintenance in the presence of more helpers. Carrion crow parents carefully monitor their chicks' begging cries – even fishing food back out of their mouths at times. This happens for all feeds, including by parents and helpers, but only parents, and usually mothers, retrieve food and eat it themselves. When biologists fed chicks, begging decreased, and parents false fed more often. This also happened when biologists then trimmed the parents' feathers to make flying harder. In contrast, helpers did not false feed, even if they were similarly handicapped, and compensated for reduced parental contributions by bringing in more food. Helpers also assist with sentinel duty, but only if they are of age.

SPECIALISED LABOUR

You might come away with the impression that carrion crow mothers care for nothing but their own welfare, but they do perform specialised tasks in addition to bearing the costs of egg laying. Adult crows are less vigilant in the presence of more adults, but ignore the presence of juveniles – probably because they provide little useful assistance in sounding the appropriate alarms. All members of a group remove faecal sacs from chicks at the nest (rather like changing nappies). Only the breeding female carries out the more specialised tasks of cleaning the chicks and the nest, and fluffing up the inner-nest layer. She probably cares the most about hygiene in the nest because she is the main incubator and brooder.

Above

Seychelles warblers are unusual among cooperatively breeding birds in that helpers at the nest are usually females rather than males.

WHO CARES?

Like most cooperatively breeding bird species, carrion crow sons are more likely to help than daughters, so groups in need of helpers tend to fledge more sons. Superb fairywren helpers are also young males, who delay dispersing and breeding to assist their parents. Daughters always leave home, which is more risky, causing superb fairywren populations to have a lot more adult males than females. In contrast, red-backed fairywrens of both sexes stay, and helpers of either sex are equally common.

Subordinates may actually be secret breeders. Male superb fairywrens may engage in furtive copulations with the breeding female, or piggyback on the attractiveness of the dominant male to gain more extra-group copulations. Subordinate female Seychelles warblers can benefit directly by sneaking their own eggs into the dominant female's clutch. Perhaps this also explains why subordinate male pied babblers tend to be ousted from a group sooner than genetic relatives of the dominant male.

Seychelles warblers are unusual among cooperatively breeding birds because most of their helpers are female. Female subordinates in this species gain three times the benefit of male subordinates. This is because females are much more likely to be closely related to the brood they are helping to rear. Daughters disperse less than sons, and only assist if their mother is the breeder, because 40 per cent of a brood can be fathered by an extra-group male. While male subordinates never get to mate with the breeding female, females sometimes sneak one of their own eggs into their mother's nest. When successful, they are helping to raise their own offspring in addition to younger siblings.

HELP WANTED

A disproportionately high number of species that must breed cooperatively live in harsh, unpredictable places, such as deserts and islands. Many of these species, including chestnut-crowned babblers in the Australian outback and sociable weavers in the Kalahari Desert, are not even territorial, so property saturation does not explain their decision to breed cooperatively.

In a natural experiment that illustrates the greater need for help when times are bad, biologists observed different helping behaviour by pied kingfishers that nest in colonies and fish at two very different Kenyan lakes. Lake Victoria provides kingfishers with skinny, nutritionally poor fish that are hard to catch, with only 24 per cent of dives being successful. These kingfishers also had a longer commute from their nests to the lake. Here, breeding pairs always had helpers, many of which were unrelated to the breeders. In contrast, Lake Naivasha provided kingfishers with plump cichlids and easy fishing (offering an 80 per cent success rate) on their doorstep. Under these comfortable conditions, helpers did little to increase the number of chicks reared, and the only helpers were sons.

HARD LIVES

The 'hard life' hypothesis proposes that when the collective benefits are sufficiently great to outweigh going it alone, individuals will invariably breed in groups. Cooperative breeding in these areas has evolved to buffer groups from unpredictable periods of drought and starvation. The hypothesis predicts that helpers are more valuable in bad times than in good, leading to more, rather than fewer, helpers in conditions with less food.

Sociable weavers, for example, nest in massive thatched nests, the equivalent of bird apartment blocks. When biologists provided these weavers with extra food, they experimentally reduced the reproductive benefits of help in dry years. White-fronted bee-eaters in Kenya, which also nest in colonies, have more helpers in dry years – when there is less insect prey and more than two adults are needed to bring in enough food. Breeders benefit more from group provisioning than in times of plenty, and 40 per cent of helpers are former breeders.

Right

Sociable weavers nest in bird condominiums, with each family adding its nest to the communal thatched structure.

BET HEDGING

The 'bet-hedging' hypothesis differs subtly from the 'hard life' hypothesis. In the former, although helpers are still beneficial in bad times, larger groups also increase reproduction in conditions of plenty. Another Kenyan bird, the superb starling, always breeds in large groups, regardless of environmental conditions. Biologists believe that as individual starlings experience such a high risk of failing to breed, due to the extremely unpredictable rainfall patterns in central Kenya, their best strategy is always to hedge their bets by breeding in groups. A stable reservoir of subordinates acts as a buffer against uncertainty. These starlings live in complex, two-tiered societies with multiple breeders.

BIGGER IS BETTER

Australian magpies are also cooperative breeders that live in large, complex groups in a harsh environment. In 2019, biologists suggested that having to navigate the relatively complex politics of a bigger group can train individual magpies to become generally more intelligent as adults. Adults from larger groups perform better in a range of IQ tests administered by biologists. It appears that body size, amount of food and other personality traits – such as being shy or risk averse – have nothing to do with this difference in intelligence. Nor is it a simple matter of the smarter birds flocking together.

However, there are benefits to being clever, as mothers with the highest test scores raise the most young. These females are also the most adept at finding food, leading to the speculation that their children get a higher quality, more varied diet, even if they do not get more calories or grow up any heavier than the offspring of mothers that tested less well. As extra-pair paternity is high among Australian magpies, we do not know how much of this IQ difference is due to genetic predispositions. However, biologists recognise that it is partly learned, because juveniles from larger groups only demonstrate a superior IQ to those growing up in smaller groups about 200 days after fledging. The intellectual gap increases after another 100 days.

Corvids, like magpies, are famously clever, but this is one of the first studies to demonstrate that living in larger groups can be one of the main things that drives wild birds to develop into more intelligent individuals.

KIDNAPPING AU PAIRS

In some cases, large groups are so crucial to survival that cooperative breeders will go to extreme lengths to ensure they have enough helpers. Southern pied babblers have been observed kidnapping neighbouring young and raising them as future helpers.

Like many cooperative breeders in harsh environments, southern pied babblers of the Kalahari Desert need help incubating eggs, brooding and provisioning chicks, escorting young, watching for danger, and defending the territory from neighbouring groups and predators.

Sequence of events during a kidnapping attempt

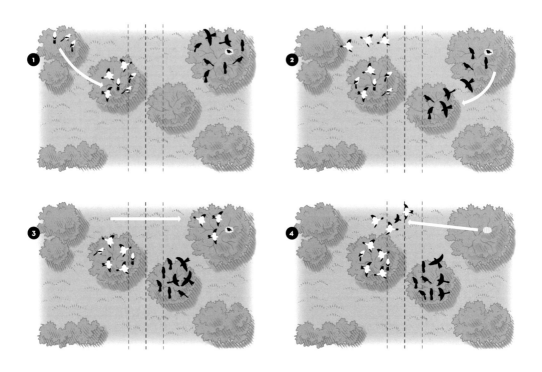

In drought years, helpers are especially useful. In such times, smaller groups or pairs generally fail to hold a territory or breed (one of the best ways to recruit help), risking extinction. These small groups have two options. They can merge, which is highly unusual in such a territorial species, creating a temporary alliance that dissolves when the drought ends. Or they can kidnap the young of neighbouring groups, and rear them as future au pairs and group members.

Kidnapping is structured around a sequence of specific events. It begins with the initiation of a border dispute (1). When the neighbours are distracted (2), the smaller group sneaks into the heart of their neighbour's territory, looking for their hidden chicks, and hoping to elude any babysitters left behind to guard the precious young (3). While holding bait in its bill, the kidnapper gives out soft food calls – a

Southern pied babblers

Southern pied babblers live in the arid Kalahari Desert, and it takes a whole group to raise a brood of chicks. Young pied babblers will respond greedily to any offering of food, even from strangers, making it quite easy for a neighbouring band of adults to lure an unsuspecting juvenile into enemy territory, where they raise it to help their own group.

combination that greedy youngsters find irresistible. As chicks will indiscriminately follow any adult holding a choice titbit, the kidnapper simply backs away until it has lured an unsuspecting victim safely across the border (4). At that point, the rest of its group abandon their posts at the decoy border dispute and join in tempting the kidnapped youngster into the heart of their own territory. Here, it is treated like one of the kidnappers' own chicks and grows up to become a helper for them.

GREAT EQUALISERS

Even though reciprocity should constrain unrelated groups to distribute reproductive benefits equally, individual members can vie for a larger share of the collective profits. Paradoxically, equality among joint nesters can be maintained through increased competition within the group.

Holding the top position in a group can be less rewarding than outsiders might think. When biologists remove a dominant male acorn woodpecker during the egg-laying period, he destroys all the eggs upon his return. This is probably because he suspects there is no way for him to have fathered any of the brood, so infanticide is the best way for him to force females to lay a new clutch that he will have some share in.

Below

Unlike most cooperative breeders, acorn woodpeckers are joint nesters, with multiple breeding pairs all sharing parenting duties and a single nest.

The dominant breeding female in white-browed sparrow-weaver groups suffers more oxidative damage (a sign of stress) than her subordinates, suggesting she is more stressed and will age faster. Among pied babblers, only one pair breeds, and infidelity is rare. However, this comes at the cost of great infighting – particularly between females in a newly formed group. In the process of vying for the position of top breeder, females will make attempts to evict one another; they will even go so far as to eat the eggs of the breeding female.

COALITIONS AND LIFETIME BENEFITS

Unlike the monogamous Florida scrub-jays with stay-at-home sons, acorn woodpecker groups consist of coalitions of both sexes, all of which breed. Most members of a same-sex coalition are relatives, so offspring that remain as helpers never mate with breeders because of a strong incest taboo. Male acorn woodpeckers are often more likely to inherit their fathers' positions after spending some time outside the group; they then return to their birthplace if a breeding vacancy occurs. When all the breeders of one sex die or leave, their place is vigorously contested by competing coalitions; the ensuing 'power struggles' can last days. Larger coalitions usually win, but no one knows why they prefer to wait, rather than initiating a takeover of a smaller, existing coalition.

LESS POLARIZATION IN POVERTY

Taiwan yuhinas form socially monogamous pairs. Both sexes establish a pecking order, and females compete to have the most eggs in the communal nest. Females tussle by landing on the back of whoever is in the act of egg laying, pecking and attempting to shove her off the nest. When there is plenty of food, dominant females tend to win the most tussles, lay their eggs earlier and initiate nocturnal incubation. The last eggs laid have the smallest chance of surviving because they hatch last, if at all.

However, when times are hard, tussles are less intense. Everyone then lays fewer eggs, resulting in greater reproductive equality between dominant and subordinate females. When there is less food to go around, subordinate females are also more likely than dominant females to have affairs with males both within and outside their groups – perhaps to improve the genetic quality of their chicks.

SORORITY SQUABBLES

Joint nesters span a spectrum of behaviour, from egalitarian to almost parasitic. Among joint nesting woodpeckers and four species of New World cuckoo, intensely wasteful conflict between cobreeding females surprisingly enforces reproductive equality. The antics of acorn woodpeckers have been described on page 149.

Opposite

A guira cuckoo egg, with its fine tracery of white over a blue base, has been compared to Wedgwood pottery.

Another woodpecker, the campo flicker, has also been observed breeding and defending a year-round territory in groups. In both species, biologists have seen eggs vanish, but no evidence of predation. When campo flickers lose a clutch to a predator, they renest in a new cavity. But these egg disappearances were never followed by nest relocation; they only occurred in the half of the population that bred in groups, rather than in pairs. The most likely explanation for the missing eggs is that females were removing one another's eggs in an attempt to lay the lion's share of a clutch.

Right

Guira cuckoos are among four species of New World cuckoo that are joint nesters.

EGG KILLING TO CHICK KILLING

Unlike the three ani species, guira cuckoo groups do not consist of socially bonded pairs. This may be because anis spend only about 1 per cent of their time on territorial defence, whereas more promiscuous guira cuckoos have no time to guard their mates.

Guira cuckoo groups contain relatives and nonrelatives; male members are more closely related than joint nesting females. Adults forage together; larger groups enhance foraging efficiency in groups of four to fifteen and anti-predator defence.

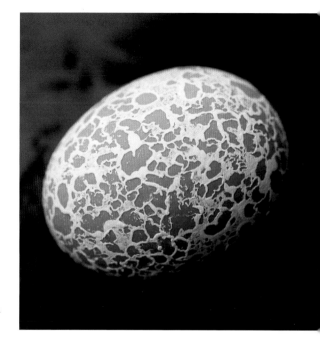

The eggs of guira cuckoos are a stunning turquoise blue, covered in a lacey pattern of the same white chalky substance that coats freshly laid ani eggs. They are so striking that an early observer compared them to Wedgwood pottery. Intriguingly, although guira cuckoos frequently eject eggs to trim down the communal clutch, these exquisite egg patterns are not signatures, and therefore cannot help females avoid killing their own offspring.

In addition to eggs, biologists noticed an almost daily disappearance of chicks, usually a few days after hatching. Some dead and wounded were discovered under nests. Others were carried 30–40 metres (98–130 feet) in an adult's bill, then dropped in mid-flight, to be attacked by several group members.

Infanticide was common, and could eliminate entire broods. No one observed any attempts to defend the chicks.

Perhaps because infanticide takes such a heavy toll, females do not engage in the anis' extravagant cycles of egg tossing. They also limit broods to two to three eggs each, regardless of the group's size. Perhaps this smaller egg investment explains why first-time layers are more likely to have no eggs in the communal nest, giving them an incentive to be infanticidal. Groups containing more relatives suffer less infanticide and conflict. Unfortunately, we may not know exactly why guira cuckoos behave this way; urban development around Brasilia has led to this study's untimely end.

DEALING
WITH DANGER

SENTRY DUTY

Living and breeding in close-knit groups has both costs and benefits. On the one hand, a larger group of birds could be more likely to attract predators or transmit diseases. On the other hand, individual birds in a coordinated group can save time and energy being vigilant. Even in completely uncoordinated groups, there is pure safety in numbers because the group dilutes the chances of any particular individual becoming lunch.

Cooperative breeders that live and function best in large groups, such as the pied babblers, do not benefit only from extra help defending a territory and caring for offspring. They save substantially on time spent in having to watch for danger. All adults in a group can act as sentries. These birds spend 95 per cent of their feeding time on the ground, probing the sand for prey. This makes it hard to keep an eye out for approaching predators, or to check the designated sentry has not gone off for a nap.

To signal that it is on active duty, a pied babbler sentinel sings a special 'watchman's song'. By playing a variety of sounds to foraging babblers, biologist Andy Radford and his colleagues have shown that the watchman's song has a specific meaning, and that it allows a group to eat more. Firstly, those feeding can afford to spend less time looking up for predators, which means they can forage more efficiently. Secondly, hearing the 'watchman's song' puts them sufficiently at ease to range farther afield. They are thus less likely to be competing for the same prey, or going over a patch that has already been depleted by another group member. There is a positive feedback loop here, in which larger groups are more likely

to have sentinels, which increases the survival of group members, which in turn leads to even larger groups.

Pied babbler sentinels and foragers can even signal their hunger levels by calling faster the hungrier they are; they then use this information to coordinate changes in sentry duties. Sentinels do not just pay the opportunity cost of not feeding: they are also more vulnerable to predators when perched in a conspicuous spot and singing constantly, enabling the rest of the group to feed in peace. However, with the exception of a few kidnapped members, pied babbler groups tend to comprise close relatives – a situation that reduces the cost of a risky and altruistic behaviour such as sentry duty. By contrast, joint nesters such as smooth-billed anis, in which group members are unrelated, offer no evidence of organised sentinel behaviour.

Among cooperatively breeding superb fairywrens, both breeding and nonbreeding males perform sentry duty. The nonbreeding helpers allow breeding males to spend more time on sentry duty than is possible for males with fewer helpers; the presence of a sentry also allows adults in the group to spend less time scanning for danger when approaching or leaving a nest of chicks. However, sentries do not actually reduce nest predation by pied currawongs, the main nest predator of superb fairywrens.

Opposite

A southern pied babbler sentinel in South Africa sounds the alarm to alert the rest of its group of possible danger.

MOBBING

You have probably encountered crows making a racket somewhere in the trees, or seen a bird of prey flying with much smaller birds in hot pursuit. You may even have been personally bombarded by gull droppings. All these are examples of a widespread behavioural defence called mobbing.

Birds mob potential threats, so mobbing is most often observed just before and during the breeding season, when reproductive hormones are surging. Early in the breeding season, red-winged blackbird males attack anything that flies through their territory, including rival males. Many species mob in an attempt to drive predators away from their nests and young. During the nesting season, Australian magpies can attack people with such ferocity (aiming for the eyes) that schoolchildren go about with buckets on their heads, and victims contribute to a live online map of magpie attacks. Mobbing also functions to blow the cover of an ambush predator such as an owl or cat, while alerting

the entire neighbourhood. Bird alarm calls are often understood across species, serving to recruit a mixed-species mob.

For social birds, mobbing can also teach younger group members what to beware of, and what to attack. Crows, magpies and jackdaws have been known to remember dangerous humans who raided their nests, and to learn who to mob by watching their elders. Konrad Lorenz (1903–89), a biologist awarded a Nobel Prize for studying animal behaviour, has some charming anecdotes in his book, *King Solomon's Ring*. There, he describes how he had to resort to dressing up in a full-body devil costume to mislead the jackdaws on his roof into believing a new person had come to kidnap their chicks.

Australian magpies are able to recognise individual people even if they change what they are wearing.

Many birds dive bomb and mob a predatory bird, especially one with the silhouette of a raptor or owl. Tyrant flycatcher species are especially aggressive, for example, eastern kingbirds (a type of tyrant flycatcher) commonly clamp their feet on the backs of hawks as part of their aerial attack on the much larger predator.

Aerial aggression against potential predators

Tyrant flycatchers such as kingbirds are particularly vehement when it comes to mobbing potential predators such as hawks.

ENCODING DANGER AND EAVESDROPPING

When birders make squeaking, clicking, or sibilant noises, called 'pishing', they are essentially trying to mimic the general alarm and recruitment calls of small birds, so as to persuade an unidentified bird lurking in the bushes to show itself. The reason this sometimes works is that many birds eavesdrop on the alarms of others.

Similarly, if biologists simulate a predator intrusion by playing mobbing calls through a loudspeaker, everyone in the neighbourhood increases their vigilance; each bird spends more time watching for danger and less time feeding. Another trick is to play the call of a perched predator, such as a small owl. In this case, too, birds tend to approach the loudspeaker or pishing human to investigate the source of possible danger.

While a few birds coordinate sentry duty within a close-knit social group, even familiar species such as the Old World tits and New World chickadees (both in the family *Paridae*) have remarkably sophisticated and specific alarm calls. These can trigger mobbing behaviour, or a very different response, depending on the threat encoded in the call.

Japanese great tits have distinct alarm calls for different types of predator. By playing back different pre-recorded calls, the biologist Toshitaka Suzuki and his colleagues found that a 'jar' sound means snake; the noise causes incubating females to fly out, leaving less scent and heat in the nest cavity for a snake to detect. In contrast, a 'chicka' call signals a predator that hunts using eyesight, such as a mammal or crow. Rather than revealing the nest location by flying out, females respond to this call by staying put, and cautiously poking their heads out of the hole to

Opposite

A Japanese tit looking for a snake in response to hearing an alarm call that signals 'snake'.

see from where the danger is approaching. There is even evidence that the tits vary the 'chicka' call structure to distinguish crows from mammalian predators.

Suzuki has conducted further experiments that involve a stick moving like a snake. These appear to reveal that the snake-specific alarm evokes a mental search image in Japanese tits. The birds only approached to mob the fake snake after hearing playbacks of the 'jar' alarm, and depending on whether the stick moved in a snake-like manner. They ignored the stick if playbacks were of other alarms, or if it did not move like a snake.

Japanese and willow tits also have calls that alert others to danger, and different calls that recruit others to help mob the threat. These two species often flock together, and seem to understand each other by the order in which the calls are made, rather than just responding to individual calls. If biologists play an 'alert' followed by a 'recruitment' call of either species, Japanese tits approach to mob. By contrast, if they hear the two calls in the reverse order, they barely bother to respond.

SOUNDING THE ALARM

Black-capped chickadees are named for their distinctive mobbing calls, 'chicka-dee-dee-dee', which they sound in response to perched predators. They also have a high-pitched, difficult to locate 'seet' alarm, which they sound if the predatory bird is flying overhead and on the hunt. To see if there were even finer degrees of meaning to the mobbing calls, ornithologists from the University of Montana teamed up with Kate Davis, a local falconer, to present wild chickadees with thirteen species of live raptor of varying size and hunting style, from owls, to hawks to falcons.

The experiment revealed that chickadees add more 'dee' notes as the level of threat increases. So, the smallest owls that specialise on ambushing small birds such as chickadees elicited the most 'dee' notes, while large, mostly mammal-eating raptors such as red-tailed hawks and great grey owls provoked the fewest 'dees'. By playing back these recordings, the biologists also showed that chickadees respond accordingly; they sound the same 'chickadee' alarm call and approach the loudspeaker most often in response to the sounds of others being alarmed by a small owl. The context-dependence of these subtle variations in mobbing calls are learned by young chickadees.

Red-breasted nuthatches often flock with chickadees. By using similar playback experiments, biologists have shown that the nuthatches understand the nuances of meaning encoded by the different number of 'dee' notes. They spend more time mobbing a speaker playing the 'chickadee' alarm that has more 'dee' notes, as well as flicking their wings (a sign of agitation), and sounding their own alarm more often.

Mixed flocks in Australia, Africa and Asia also comprise species of birds that eavesdrop on the alarm calls of others. Superb fairywrens and white-browed scrubwrens understand the degree of threat encoded in each other's alarm calls. African fork-tailed drongos that deceptively sound the alarm to steal food caught by credulous pied babblers are generally trusted sentinels for other species that spend more time on the ground. Similarly, the greater racket-tailed drongos of Asia are trusted sentinels among mixed-species flocks in the Sri Lankan forests.

Mammals and birds also eavesdrop on one another. Burrowing owls become extra alert when they hear prairie dog alarm calls, vervet monkeys respond to superb starling alarms, dwarf mongooses know hornbill alarm calls, and red squirrels take cover when Eurasian jays sound the alarm. Eastern grey squirrels do not just respond to bird alarms for a red-tailed hawk, but also relax more quickly when hearing the contented chatter of feeding birds than silence.

A danger scale

The larger a predatory bird, the less able it is to manoeuvre adroitly in midair to catch a small songbird. Black-capped chickadees add more 'dee' notes to their alarm calls when they encounter the most dangerous predators, which, in this case, are Northern pygmy owls (1) and Cooper's hawks (2), which specialise in hunting small birds. Peregrine falcons (3) are less threatening as they tend to hunt larger birds in the open, and least threatening of all are large, mammal specialists such as red-tailed hawks (4) and great grey owls (5).

REPRODUCTIVE CHEATS

In addition to sounding the alarm because of predators, small songbirds such as Japanese tits or reed warblers benefit from alerting each other to the presence of a brood parasite, for example a cuckoo. This makes a lot of sense – the best way to avoid the costly mistake of raising a foundling is to prevent its mother from laying her eggs in your nest. If a common cuckoo succeeds in laying her egg, and host parents are unable to remove it, the cuckoo chick proceeds to kill all their own biological offspring. It will also exhaust its host parents, as they struggle to feed a chick that grows to be several times their size.

BIRDS THAT PARASITIZE OTHER SPECIES

- Cuckoos (evolved three times independently)
- Cowbirds
- *Vidua* finches
- Honeyguides
- A single duck species, the black-headed duck

In Europe, reed warblers learn to recognise and mob common cuckoos by watching their neighbours; they then increase their overall vigilance when they see and hear their neighbours attacking cuckoos. In China, neighbouring hosts of the common cuckoo learn to understand and eavesdrop on the alarm calls of a different species. Watching oriental reed-warblers mobbing cuckoos seems to be enough for neighbouring black-browed reed warblers to approach when they hear the alarm calls of the other species.

In response to these mobbing attacks, common cuckoos have evolved to mimic sparrowhawks, a common predator of small birds such as reed warblers. Female common cuckoos come in two colour morphs. Grey cuckoos look so much like sparrowhawks that reed warblers have to think twice before attacking them; they only learn to mob this morph by watching their neighbours doing so. The red cuckoo morph is very rare – making it harder for hosts to encounter them often enough to learn that they, too, should be mobbed if seen near a nest.

A more common way for brood parasites to minimise being attacked is to lay their eggs so fast and stealthily that the host parents do not even notice the intrusion. I vividly remember waiting with bated breath at a barbet nest hole in Zambia, while a lesser honeyguide female approached. She waited until both barbets were out, and then popped into the nest hole for two seconds. Then she was out, with an egg in her bill and, presumably, her own freshly laid in the nest.

INCREASING EGGSPERTISE

Experience also leads to better egg discrimination. Hosts with the ability to reject parasite eggs do so not by spotting the odd egg in a nest, but by learning what their own eggs look like and rejecting any that look sufficiently different. This also selects for increasingly effective egg forgeries by brood parasites, making it harder and harder for hosts to know which egg to remove. Common cuckoos have been locked in an arms race of egg recognition and mimicry with some host species for so long that lineages called gens have evolved to specialise on particular host species. By contrast, other brood parasites are generalists, and may be at a much earlier stage of the evolutionary arms race. The brown-headed cowbird parasitizes about 140 different host species, and there is less evidence of specialisation by particular females.

Right

Clutches of different songbirds' eggs, in each clutch is a mimicking cuckoo egg.

Left

A bee-eater clutch parasitized by a honeyguide; the largest egg is a honeyguide egg, and the smaller eggs are all rotten because they have been punctured by the laying female honeyguides.

In a rather fitting twist, parasites can parasitize each other when multiple females lay in a single host nest. This can result in some benefits to the parasites. For instance, having more than one parasite egg in the nest can make it harder for hosts to reject the right egg. However, it can also result in an escalating arms race between females of the same parasite species. For instance, greater honeyguides parasitize hole-nesting species, in which everything is so dark that both hosts and parasites lay white eggs. Female honeyguides from different races lay eggs that closely resemble their host's eggs in size and shape. Surprisingly, their most common hosts show no egg rejection abilities and would sit on anything put in their burrow. Yet other female honeyguides parasitizing the same nest would always stab the most dissimilar eggs. In this system it appears to be the female parasites, rather than their hosts, that are driving the evolution of egg mimicry within each specialised race.

Similarly, little bronze cuckoos of Australasia lay dark, cryptic eggs that are hard to spot within the dark, domed nests of their hosts. While these hosts show no egg rejection abilities, even of experimental eggs manipulated to look shinier and more visible, female cuckoos parasitizing the same nest would selectively pick out and destroy the most visible eggs before laying their own. These experiments again suggest that competition between female cuckoos for the same host nest, rather than egg rejection by hosts, is the driver behind cryptic eggs.

DEGREES OF FLEXIBILITY

The chicks of brood parasites can flexibly tune their behaviour to increase survival in a host nest. Cowbird chicks, for example, have evolved forebrain circuitry that enables them to tune their begging calls flexibly, thus manipulating any host species to feed them to the maximum.

Common cuckoo chicks from a reed warbler specialist race have the innate ability to recognise reed warbler alarm calls; they soon learn to stop begging for food when they hear it, so as not to attract a predator's attention. Interestingly, they fail to learn to keep quiet when experimentally cross-fostered to other host species with different alarm calls. This indicates more specialisation and less flexibility than in the generalist cowbird chicks.

Mafia enforcement, in which brood parasites destroy the nests of hosts that rejected their eggs, is a thrilling example of flexible behaviour. It implies that female parasites remember which nests they have parasitized and then monitor them, selectively punishing hosts that reject the parasite's eggs. Eurasian magpies learn to accept greater spotted cuckoo eggs in response to these punishing attacks. When biologists consistently play cuckoo calls and put out stuffed cuckoos to simulate areas of high cuckoo density, the magpies will even choose to nest farther away, perceiving such places to be at high risk of parasitism.

Female cowbirds have special mental adaptations for parasitism that allow them to be extra flexible. They have better spatial memories than male cowbirds, enabling them to monitor nests better, remember which hosts raised more cowbird chicks, and target those both in the same year and in subsequent years. Brown-headed cowbirds even 'farm' their hosts by destroying nests that are too advanced for them to parasitize, thus inducing the hosts to start all over again.

Brown-headed cowbirds pose a conservation threat to many small songbirds in eastern North America. This is because, unlike hosts that coevolved with cowbirds on the Great Plains, naive hosts are comparatively defenceless against parasitism.

The Kirtland's warbler, a recent and naive host of the brown-headed cowbird, almost went extinct due to a combination of habitat destruction and parasitism. To help conserve these warblers, one of the conservation measures involves culling cowbirds. Placing a few 'Judas cowbirds' in an enclosure, with a one-way entrance to attract passing individuals, easily traps the gregarious cowbirds, which are then killed. The warbler population has now recovered

so successfully that the species may be delisted from the Endangered Species Act. Towns in northern Michigan hold annual Kirtland's warbler festivals to celebrate this rare species.

Above

A male Kirtland's warbler perches on a pine in Michigan. Kirtland's warblers almost went extinct, partly because of parasitism by brown-headed cowbirds.

SAFETY STARTS AT HOME

Nesting birds are vulnerable to dangers other than brood parasites. Parent birds thus go to all sorts of lengths to keep their nests safe, from choosing a safe location to camouflaging the nest. Others actively attack or distract potential predators.

Some offspring are more independent than others. Waterfowl hatchlings have to be extra hardy because of the inaccessible (and safe) places in which their parents choose to nest. Newly hatched wood ducks launch themselves from tree holes as high as 18 metres (60 feet) above the ground. Barnacle goose hatchlings have to rely on their fluffy feathers to parachute down 122 metres (400 feet) of sheer cliff to find food. By contrast, the pink, bald and helpless altricial songbird chicks are like the popcorn of the forest; even herbivores such as deer and moose will gladly eat the packets of protein and fat in a songbird nest.

Murrelets are diminutive members of the auk family, a group of fish-eating, diving seabirds. However, marbled murrelets fly many kilometres inland to nest in old-growth conifers in the Pacific Northwest, while Kittlitz's murrelets nest near glaciers on inland mountaintops in Alaska. There are fewer competitors and predators in these remote places, but no one knows exactly how their fledglings – which are too covered in down to fly efficiently – get all the way back to the coast. This species' highly specialised choice of nesting and feeding sites has made it particularly vulnerable to climate change and oil spills.

Hoatzins – also called stinkbirds because of the smell emanating from the leaf-digesting bacterial chambers in their stomachs – prefer to nest over water in the Amazon Rainforest. These flying compost heaps have been on their

Above

Wood ducks nest high off the ground in tree holes, and their precocial ducklings launch off soon after hatching, bouncing like thistledown on the forest floor before following their parents to food and water.

Opposite

In stark contrast to the independent, downy young of ducks and chickens, songbird chicks like these begging blackbirds are altricial, remaining bald, pink and relatively helpless for days after hatching.

own evolutionary trajectory for 65 million years: their chicks resemble archaeopteryx and other birdlike dinosaurs in having wing claws. Although the claws disappear in adulthood, the chicks make good use of them. The youngsters drop into the water when predators approach the nest, and then use their wing claws to clamber back out of the water.

Some tube-nosed seabirds, such as storm petrels, return to their burrows under cover of darkness, probably to avoid being eaten, or having food meant for their chicks stolen by larger seabirds such as gulls and skuas. Curious about how these birds found their way home in a colony without even a glimmer of moonlight, biologists kidnapped adults and popped them in a two-armed maze: one arm led to their own burrow, the other to that of a neighbour. Both burrows were uninhabited during the experiment. The birds invariably went down the arm leading to their own tunnel. However, when their nostrils were injected with a zinc solution, which temporarily blocked their sense of smell, the birds were unable to distinguish between the two tunnels. Across species of petrel, only those that navigate home in darkness use their sense of smell, while those that return during the day rely on their vision.

DECEPTION AND DEFENCE

Opposite

The ingenious common tailorbird uses plant fibres or spider silk to sew a large leaf together to create a well-camouflaged nest.

Below

A paraque, a species of nightjar with its chick, both well-camouflaged on their ground nest.

Camouflage is an effective alternative to nesting in inaccessible or remote places. Ground nesters, such as quail and nightjars, lay eggs that blend in so perfectly with their surroundings that I have almost trodden on them. Members of the *Cisticola* family, classic little brown birds found in Eurasia and Africa, use silk to contrive some beautifully camouflaged nests. One of my favourites is the rattling cisticola. This bird sticks growing grass together with concentric circles of silk on the inside of a nest that ends up resembling a narrow vase. These nests were so perfectly camouflaged that we needed the help of many sharp-eyed young assistants to find them in the grasslands; local children referred to them as 'bottle' because of their resemblance to a soda bottle. Tawny-flanked prinias stitch a pouch of woven plant fibres to green leaves, using silk as thread.

Plovers such as killdeer also have beautifully camouflaged eggs, but these birds go a step further by performing distraction displays. In these, a parent lures an approaching predator far away from its eggs or chicks by flopping about helplessly, mimicking an easy target with a broken wing. Other ground nesters, including some members of the chicken order, for example willow ptarmigan and red grouse, also perform distraction displays. Even more impressive are brown thornbills, which mimic a chorus of alarm calls by other species to deceive their main nest predator, the pied currawong, into retreating at the perceived threat of a predatory hawk.

Counterintuitively, some small birds use large and powerful neighbours as protection. Black-chinned hummingbirds that nest near northern goshawks and Cooper's hawks lose fewer offspring to marauding Mexican jays. Both hawk species commonly prey on other birds, especially those the size of a juicy jay, so Mexican jays avoid foraging near the hawk nests. This effectively creates an enemy-free space for the diminutive hummingbirds – presumably not worth a hawk's while.

Other birds actively threaten or attack predators. Several species of tits hiss when a predator approaches the nest. Willow, great, and blue tits suffer about 20 per cent less nest predation by yellow-necked mice than another small cavity-nester, the collared flycatcher, which also nests in the same area. Using playback experiments, biologists in

Poland found that while mice visited all nest cavities equally often, they spent an average of less than 4 seconds investigating nest boxes playing hissing sounds from any of the three tit species, compared with more than 26 seconds in silent boxes.

The greater roadrunner, a species of non-parasitic cuckoo in the southwestern US, loses over 70 per cent of nests to predators such as snakes and coyotes. About half the parent roadrunners attack predatory rat snakes that are invading their nests by striking the snakes repeatedly with their large, sharp bills. These defences saved at least one egg or chick in more than half the nests with proactively protective parents.

MATERNAL INVESTMENTS

Mothers can, and often do, vary their reproductive investments, depending on how dangerous the environment is. Just perceiving an increased risk of nest predation is enough to make a female lay fewer eggs.

Below

A Eurasian sparrowhawk catches a coal tit in mid flight. Sparrowhawks are specialist bird hunters, and are adept at manoeuvring in midair to catch small birds.

When biologists play pre-recorded predator calls to simulate a higher risk of predation, female song sparrows respond by laying fewer eggs, nesting in thornier vegetation, and modifying nest behaviour to attract less attention from predators. The females also spend shorter bouts incubating

eggs, more time off the nest, and visit less often to feed their chicks. The apparent threat of increased predation is enough to reduce the reproductive success of a nest by 40 per cent. This reduced breeding success occurred specifically in response to hearing the calls of predators such as crows, hawks, owls and raccoons; sparrows that listened to harmless sounds such as seals, geese, hummingbird or woodpecker calls raised just as many chicks as they would normally do.

MOTHERS UNDER STRESS

Female great tits exposed to stuffed predatory Eurasian sparrowhawks while laying eggs produced lighter chicks with disproportionately larger wings, presumably to aid escape. Just injecting European starling embryos with stress hormones is enough for chicks to grow up with stronger flight muscles and to be more powerful fliers. Among yellow-legged gulls, mothers exposed to stuffed mink by biologists perceive a higher predation risk; they programme anti-predator reflexes in their chicks, probably by laying eggs with more stress hormones. These chicks with stressed mothers who had spent time mobbing mink responded faster to adult alarm calls by crouching and staying still.

Even more impressively, these yellow-legged gull chicks warn each other of danger through the eggshell. Biologists exposed some eggs to constant recorded playbacks of adult gulls screaming at predators, then placed some of these 'experienced' eggs alongside others that had heard only background colony noise. They found that both sets of eggs vibrated more than eggs never exposed to alarm calls either directly or indirectly through their neighbours.

Furthermore, the eggs that vibrated more had higher stress hormone levels; they hatched into chicks that crouched faster in response to alarm calls, and made fewer begging noises. The downside of the stress is that these same predator-prepped chicks grew more slowly, and fledged at a lighter weight than their counterparts with a more secure experience in the egg.

LIVING DANGEROUSLY

Role-reversed species, such as jacanas and spotted sandpipers, often nest dangerously, where there are predators as well as abundant food. This high-risk, high-reward strategy could help explain why females turn into egg-laying machines to replace clutches lost to predation, while males can manage as single parents of their chicks. Jacana females defend sought-after territories with enough food to support multiple males, and in an area compact enough for one female to monopolise. Unlike spotted sandpipers, they do not help with nesting or incubating eggs. However, like spotted sandpipers, jacanas are simultaneously polyandrous, easily revisiting any male that has lost his clutch to predators.

STRATEGIC PARENTING

In addition to preparing chicks for a riskier world before hatching, parents often vary behaviour to minimise risk from predators. Some of these risk-related behaviours distinguish species with very different environments and life histories; others reflect the immediate environment, or an individual's personality.

Across 182 songbird species, those in the southern hemisphere (Australia, New Zealand and South Africa) have slower life histories than relatives in the northern hemisphere (Europe and North America). In other words, southern-hemisphere species tend to live longer, and literally put fewer eggs into a single reproductive basket compared to closely related species of a similar size in the northern hemisphere. These reproductive attempts also play out in parental risk strategies.

Biologists have compared matched species pairs in Argentina and Arizona in an experiment. They explored how parent birds responded to taxidermic mounts of predators, accompanied by vocalisations played through a speaker. All parents exposed to fake predators cut back on risky feeding visits to their nests – a sensible strategy, as these could attract predators or reveal the nest. Yet the longer-lived species with smaller clutches reduced the number of risky nest visits more in response to sharp-shinned hawks, a common predator of adults; in contrast, the species with larger broods responded more strongly to

presentations of Steller's jays, a common nest predator. In other words, longer-lived parents acted to reduce risk to themselves rather than their offspring, while species with shorter life expectancies and more offspring per brood minimised the risk to their young rather than to themselves.

PROTECTIVE PARTNERSHIPS

In about three-quarters of all bird species, both sexes contribute to parental care. Parental coordination can happen in several ways, but is one of the main reasons for pair-bonding rituals among socially monogamous birds. Superb fairywrens, for example, respond more strongly to the alarm cries of their mate than to the alarm cries of another individual of the same species; they wait longer before visiting the nest, and join in the alarming faster. Among blackcap warblers and long-tailed tits, pairs that more seamlessly synchronise feeding visits to the nest are less likely to lose chicks to predators. This is probably because behaving like a well-oiled relay team reduces the amount of noise and fuss at a nest every time an adult lands, meaning that predators are less likely to be alerted to a nest.

Biologists have found a surprising variation in the lengths of individual incubation shifts among thirty-two shorebird species that share parental duties with their mates. The best explanation for this variation was how parents reduced predation risk. Species that relied on camouflage, such as long-billed dowitchers, incubated for 50 hours at a stretch, thus minimising activity at the nest that could attract predators. In contrast, species such as ringed plovers, with distraction displays, or those who left the nest well before a predator approached in response to their mate's alarm call, could afford to be more active around the nest.

SAFETY IN NUMBERS

Anyone watching a school of fish or a flock of birds twisting and turning in breathtaking synchrony will wonder how thousands of individuals coordinate their movements so precisely with no evidence of a conductor or commander.

Collective behaviours such as flocking are the emergent properties of a few simple, individual rules. Computer simulations show that individuals regulating their response to their closest neighbours are enough to produce the complex and coordinated flight patterns of a massive group. Exactly how the responses are perpetuated so fast through an extended network remains something of a mystery. However, not all individuals or positions in a flock are identical. In addition to using information on the distance from, and direction of, their neighbours, starlings use their location in a flock to tune their behaviour in a murmuration. The flock projects a shadow onto each bird's retina, telling it if it is in the middle or the edge of the group.

Even more striking are the effects of social bonds on flocking behaviour. Jackdaws mate for life, and pair-bonded birds follow different flocking rules from birds with no partner. Paired birds fly as though joined by an invisible spring, accelerating more when farther away from their mate. This allows them to fly at a relatively constant distance from their partner; pairs thus flap less than unpaired birds, saving energy.

Paired jackdaws respond to about half the number of neighbours as unpaired birds, which, like starlings, respond to seven of their nearest neighbours. This lack of connectivity makes information transfer through the flock less efficient as the proportion of mated pairs increases. In other words, jackdaws in bonded pairs save on individual energy costs, but at the higher collective price of a less efficient response to threats such as predators.

Jackdaws also follow different flocking rules in different situations. When flying to roost, they stick to a constant number of neighbours, enabling the flock to remain equally coordinated regardless of density. By contrast, jackdaws gathering to attack a predator pay more attention to their distance from their neighbours in the flock than to the number of neighbours. This results in disorganised flocks when individuals are spread out, and much more synchronous flight when everyone is bunched tightly together.

Separation
Steer to avoid crowding neighbours

Alignment
Steer towards the average direction of local flock mates

Cohesion
Move towards nearby flock mates

Collective behaviour of a flock in flight

Studies of European starlings have found that each individual responds to about seven neighbours. Thousands of individual birds – following the same simple rules of being neither too close nor too distant from their neighbours, and orienting in the same direction – produce the highly synchronised collective behaviour of a flock in flight.

GEOMETRY FOR THE SELFISH HERD

Large aggregations are predator magnets, and it is not uncommon to see a cloud of birds suddenly bending and twisting in response to a dive-bombing falcon. However, there is also safety in numbers, because each individual stands a smaller chance of being killed in a larger group. This dilution effect gives rise to groups large enough to attract more predators, simply as a by-product of individual self-interest. The tightly knit nature of flocks, swarms and schools is also a by-product of the fact that the safest spot is in the middle of the group.

V-FORMATION FLYING

Fighter jets fly in V formation to save fuel. Migrating geese and other large birds have done the same for millennia. However, aeroplanes do not flap their wings. The key to saving energy for large birds is to be able to precisely coordinate their wing beats to catch the uplift created by the bird in front. Individual bald ibis do this while positioning themselves, very precisely, just to the side of the bird in front. If they are temporarily stuck directly behind another bird, they reverse the timing of their flaps to minimise the downdrafts of air flowing off the body in front.

Above

An aerial view of a great cormorant colony nesting on dead trees in Latvia.

Right

In another example of bird colonies leaving a physical imprint, great cormorant guano (excrement) is so acidic that it kills all the vegetation underneath. This allows people to distinguish these rookeries from those of herons and egrets, even at a distance.

THE SELFISH CITY

Although many species find safety in numbers, living in crowds does have a price. The extinct passenger pigeon is a cautionary tale much touted by conservationists, eager to demonstrate humans' ability to take vast numbers of a species for granted.

These North American pigeons were hunted to extinction, because their strategy of moving and breeding in the hundreds of millions misled people into assuming these birds a limitless resource. We now think this species could only breed in vast colonies (known to colonial Americans as cities) so as to swamp predator populations. This is similar to the way in which wildebeest calve within days of each other, and periodic cicadas emerge *en masse* every 17 years.

Group courtship can dilute the risk of being picked off for each individual, but a larger collective is also more likely to attract predators, especially if lekking males also use sounds to attract females to the competition arena. I was once lured in by the sound of turkey cocks gobbling – only to find that the sounds were recordings, when the hunter playing them emerged from hiding and pointed a gun at me.

Another problem that often besets large, dense groups is the rapid spread of disease. In an especially encouraging example of how citizen science can help conservation, biologists from the Cornell Lab of Ornithology have used data from volunteers to track the spread of an epidemic of house finch conjunctivitis. Information comes from annual Christmas bird counts conducted all over the US, as well as monthly updates from 10,000 volunteers with bird feeders. This has allowed biologists to track the spread of the disease from its outbreak in Maryland in 1994 to its westward spread into California by 2009.

Ironically, bird feeders are also a primary means of increasing the transmission of this highly infectious disease, simply because birds flock to them in relatively high densities. However, the data from feeders has allowed biologists to understand how this disease coevolves with its bird hosts, and how house finch populations are able to evolve resistance. Some of their findings can be generalised and applied to how other diseases, including those affecting humans and livestock, spread and evolve. In addition, we now know that it is important to keep feeders as widely spaced as possible and to clean them regularly. This minimises feeders becoming infection hubs, which can also affect other bird species that visit them.

DANGER FROM HUMANS

Humans and other animals, such as domestic cats, can threaten birds. However, some species have learned to adapt in surprising ways to the environmental harm caused by people.

In addition to spreading disease, bird feeders can cause some bird species to decline by attracting predators, thereby changing an ecosystem's composition. Across neighbourhoods in Columbus, Ohio (US), supplemental food in the form of bird feeders drew more American crows – common nest predators. Consequently, neighbourhoods rich in bird feeders and crows saw less than 1 per cent of American robin nests fledge, compared with a fledging success of over one-third in neighbourhoods without feeders. Northern cardinals, however, seemed unaffected by the increased predators. This may be because, unlike American robins, they are seedeaters, so the benefits of food aid offset the costs of increased nest predation.

GROWING PAINS

Songbirds fledglings that have left the nest spend several days climbing about or falling to the ground as they practise flying, providing easy prey for predators such as domestic cats. Most songbird fledglings leave the nest early in the morning. By monitoring hundreds of nests from seventeen species in Illinois that nest in shrubs, biologists have shown that chicks leave riskier nests earlier in life and earlier in the morning. These

could be adaptations to minimise the perilous time spent in a nest closer to the ground or more conspicuous, but a fledgling on the ground in the early morning is still vulnerable.

NOVEL PREDATORS

In the lower forty-eight states of America alone, free-ranging cats kill up to 4 billion birds a year. Birds also take much longer to start using a new bird feeder if there are free-ranging cats about. Cats, rats and other species associated with humans have been particularly tricky for many island bird species – not just for flightless ones, although those have been especially hard hit.

In New Zealand, birds evolved with no mammalian predators until the last 700 years. Yet in this time one of the native songbirds, the New Zealand bellbird, has evolved parenting strategies that reduce predation risk. Bellbirds from an island never invaded by mammals perform all sorts of risky parenting behaviours; they engage in plenty of activity around the nest, trading off incubation duties and feeding chicks frequently. By contrast, bellbirds exposed to mammalian predators have evolved to be

more risk-averse, and to minimise telltale movements near the nest. This applies even in a bellbird population from which all mammals were removed three years ago, in a conservation effort. The anti-predator parenting strategies seem to be partly hardwired, not just a flexible response to the presence of predators. Some birds evolve fairly rapidly in response to novel predators.

Above

Fledging the nest is dangerous. Most fledglings get a head start in life by graduating from the nest heavier than their parents, but this makes them especially attractive to predators. Oilbirds nest colonially in caves in South America, and navigate by echolocating. Fledglings, up to one-third heavier than their parents, were historically harvested and boiled down for oil, earning the species its common name.

ISLAND BIRDS

In a particularly endearing example of how easily genetically hardwired behaviours can evolve in small populations, conservationists inadvertently bred for birds with a singularly maladaptive habit. The black or Chatham Island robin is a species found only on a few islands off the east coast of New Zealand. By the 1980s, introduced predators had reduced this species to a single breeding female lovingly named 'Old Blue' and her two or three consorts. Conservationists began a frantic breeding programme, fostering Old Blue's children and grandchildren to other species so they could induce the black robins to lay more eggs. This worked admirably – except for the unfortunate habit that half the females (all descended from Old Blue), had of laying their eggs on the nest rim, where they would invariably fail to thrive.

Part of the initial efforts to save this species involved having teams of people carefully monitoring every nest and replacing the mislaid eggs inside the nests, where they would be safer. When there were enough birds to construct a pedigree, biologists found that a single dominant gene was enough to make a female lay her eggs on the rim instead of inside the nest. This discovery put a stop to any misguided attempts to help the females who misplaced their eggs – enabling natural selection rapidly to remove that gene and its unfortunate consequences from the population. There are now 250 black robins, and all the females seem to lay their eggs in the right place.

The Hawaiian duck and other endangered waterfowl were stricken with a paralytic avian botulism that is found worldwide, but almost pushed these rare species into extinction. The disease is most easily spread to other dabbling ducks, likely to eat infected maggots that hatch out of an infected duck corpse. In this case, Labrador retrievers trained to sniff out the offending corpses have saved the day for the Hawaiian duck.

THIEVES AND TRAPPERS

Egg collecting without a scientific permit is illegal, but the passion still exists among some collectors. An infamous thief stole so many eggs from protected sites and species that he was convicted six times, had his collection of over 2,000 eggs confiscated, and was put on a special 'egg collector watchlist'. To prevent egg thefts, British bird lovers from organisations such as the Royal Society for the Protection of Birds take turns guarding the nests of rare species round the clock.

Other threats persist, however, such as the use of mist nets to harvest millions of migrating songbirds in Mediterranean countries, where they are considered delicacies.

Below

Like many island species, black robins, found exclusively on the Chatham Islands, about 800 km (500 miles) from New Zealand, almost went extinct.

POLLUTION

Noise pollution from roads and oil fields has disrupted the courtship leks of endangered sage grouse. Males that encounter more noise from human activities have higher stress levels, and leks are dwindling because fewer males gather to display at these noisy sites.

Ospreys have a fondness for the synthetic baling twine that ranchers use to tie hay bales together, and frequently tie knots in the twine to hold their large nests together. Over the years, nests festooned with these synthetic materials have become traps for both adults and chicks, which become fatally ensnared in lengths of knotted twine. Biologists and ranchers in Montana have begun a movement to pick up the twine rather than leaving it strewn across fields, and are giving nests a 'haircut' to prevent future entanglements.

The large eggs of birds of prey, such as bald eagles and peregrine falcons, were especially susceptible to the effects of DDT. Like many toxins, DDT – a popular pesticide in the mid-1900s – accumulates as one moves up the food chain. This had an especially nasty effect on raptors. It interfered with their ability to deposit calcium in their eggshells, resulting in many eggs quite literally breaking at the touch of a feather. The eminent conservationist Rachel Carson (1907–64) was instrumental in raising awareness about the ecological impact of DDT, and it was banned in 1972.

Systemic agricultural pesticides called neonicotinoids, which have caused many species of bees to decline, also make it harder for birds to put on weight before migration. In Germany, the insects many birds feed on have declined by almost 80 per cent in 40 years. In France, agricultural birds have been the hardest hit, declining by one-third in 17 years. Even common species such as meadow pipits have suffered population drops of 68 per cent.

POISONS

Common chemicals toxic to birds:

- DDT
- Neonicotinoids
- Diclofenac
- Microplastics

HABITAT LOSS

Grassland birds worldwide have been especially hard hit by agriculture – not just from pesticide use, but also from habitat destruction due to extensive monocultures. In 2019, scientists reported a 29 per cent drop in the number of North American birds compared to 50 years ago. Grassland species have suffered the biggest declines, accounting for 717 million of the total 2.9 billion birds lost. In tropical forests, large-scale agriculture such as oil palm or coffee plantations has also taken its toll on bird diversity.

These estimates would have been impossible without decades of records amassed by birdwatchers. Bird enthusiasts have also, very occasionally, helped to rediscover species. A Jerdon's babbler, last recorded in 1941, was spotted in Myanmar in 2015, and in 2007 an Indonesian Banggai crow was seen for the first time in 107 years. A review paper in 2011 reported 144 birds not to be extinct after all – most rediscoveries happened after 1980.

Below

Ospreys often use baling twine as nesting material, which can have deadly consequences if adults or chicks become entangled.

COPING
WITH CLIMATE

KEEPING UP WITH CLIMATE CHANGE

Below

A great tit returning to its nest hole with a caterpillar for its chicks. In spite of climate change, these birds successfully time their clutches to coincide with a peak in caterpillar food.

Climate change is arguably the biggest threat facing not just birds, but all species, humans included. The question is not whether populations can adapt to environmental uncertainty, but if they can respond fast enough, and in ways that avoid an ecological trap.

In a recent example of an ecological trap, biologists from Finland reported in 2018 that ground-nesting farmland birds, such as Northern lapwings and Eurasian curlews, have adapted faster than farmers to progressively earlier springs. In the past, farmers would have sown their seeds by the time the birds started nesting. Now, the majority of nests are plowed because the birds are nesting so much earlier than before.

In a more encouraging example, American kestrels appear to have advanced their breeding date by about two weeks to match the earlier sowing of crops in some parts of the American Midwest.

To what extent can birds adjust their breeding time to optimally coincide with food and minimise danger in order to raise as many offspring as possible? To answer this, we return to the great tits of Wytham Woods in Oxford (see pages 41, 42), where every individual bird has been closely monitored for over 45 years. Great tits rely primarily on winter moth caterpillars to feed their nestlings, and parents that miss the peak caterpillar glut raise fewer young than those that get the timing right. Over years, the caterpillars have been emerging earlier to capitalise on new oak leaves, which have also been leafing out earlier in response to untimely springs.

Across their range in Europe, the oaks, caterpillars and tits seem to get the timing right by responding to warming spring temperatures, but even without climate change, there can be up to three weeks of variation across years. There is also variation at a much smaller scale, such as within Wytham Woods. Biologists have found that across the last 45 years, the great tits have been able to synchronise their laying and first hatch dates with their food sources. To do this, they don't simply use a global cue such as temperature, but also fine-tune the dates to match the foliage of oaks within 20 metres (66 feet) of their nests. The upshot is a very subtle patchwork of tit families within a single wood, all developing at slightly different times to match the small variations in oak growth around each nest. This fine level of adjustment could be helpful for birds adapting to larger changes in climate.

However, we know from earlier chapters that food is not the only element influencing when birds breed. Great tits also vary their laying dates and incubation behaviour to minimise predation. European sparrowhawks specialise in hunting small birds such as tits, and also time their breeding to coincide with their prey, so that there are plenty of fat, vulnerable fledglings to feed to their own growing chicks.

FLEXIBLE TIMING

Individual great tits vary in their responses to peak predation risk from sparrowhawk parents, and these different strategies are part of inherited personality syndromes. Great tits with fast-exploring personalities tend to move around more, and are at a much higher risk of being killed by sparrowhawks than their more risk-averse neighbours. When biologists simulate higher predation threat by playing sparrowhawk calls, these more at-risk personalities lose more weight than their relatively stolid counterparts, as lighter birds are better at dodging a sparrowhawk.

Similarly, these fast-exploring females were more likely to lay eggs earlier when they thought the risk of sparrowhawk predation was high (from hearing experimental playbacks of sparrowhawk calls). This would leave the more cautious individuals to bear the brunt of peak predation levels when the sparrowhawks had hungry chicks to feed. As these personality specific responses are partially inherited and the result of natural selection, biologists are hopeful that great tits can evolve to be even more flexible should the environment become more unpredictable in the future.

A GLOBAL PROBLEM

Most species that breed in temperate regions are under strong selection to breed earlier due to climate change, and have successfully done so by arriving at the breeding grounds and laying eggs earlier. A 2019 study summarising the findings of almost sixty research papers estimated that although most bird species and populations have been keeping up with climate change by breeding earlier, not all will adapt enough to escape extinction. In particular, great reed warblers and snow buntings are at high risk, whereas innately flexible species such as great tits, European magpies and song sparrows have a better chance of persisting in the face of shifting seasons.

Opposite

A Hudsonian Godwit in breeding plumage; these birds have a long distance to migrate, and risk arriving after the insect hatch has peaked, as spring temperatures soar and advance with the changing climate.

Birds that breed in the Arctic, such as Hudsonian Godwits, are especially vulnerable to climate change because of the comparatively narrow time span in which they have to breed, and the exceptionally long distances that they migrate. Red knots, another long-distance migrant, have shrunk by 15 per cent since 1985. This could be because the Arctic spring has advanced by about two weeks, and the birds arrive too late to catch the peak of hatching insects, and their young families are raised on less food.

About seventy shorebird species worldwide migrate between the top and bottom of the globe every year. It remains to be seen if any of these populations will adapt to the changing timing of the food gluts they rely on to rear their families.

In a study based on data reported on the citizen science platform eBird, biologists estimate that for seventy-seven species that migrate within the Americas, almost all will encounter novel climates throughout the year. Not just in their temperate breeding grounds or tropical wintering grounds, but throughout migration as well, particularly when juveniles are about to embark on their first journey south. These studies focused primarily on northern hemisphere birds, so the jury is still out on how species from the tropics or southern hemisphere are likely to respond to climate change.

CLIMATE AND COURTSHIP

Changing climates can disrupt courtship and sexual selection by creating a mismatch between sexually selected signals and the quality they originally reflected. It can also alter species ranges, which changes some of the boundaries between closely related species that occasionally interbreed.

Opposite

A mating pair of rock ptarmigan. Males moult into a camouflaged brown later than females so as to attract a mate with their conspicuous white plumage.

European barn swallow nests have become smaller as the breeding seasons have become warmer. We know that males with longer tails tend to make smaller nests, and that genetic variation underlies differences in male tail length. The upshot of climate change is that female barn swallows in Europe have selected for males with even longer tails than in the past. With warmer springs, females no longer need to settle for a less sexy male in order to have a bigger nest.

Climate change can also disrupt how birds trade off investment in sexually selected signals of attraction as opposed to parental care. Hume's warblers breed in the western Himalayas, and have adjusted to earlier springs by advancing their breeding season by two weeks. Both sexes use the size of their yellow wing bars to choose mates; and birds that hatched earlier tend to develop larger wing bars, because of a greater investment by their parents. So, from 1985 to 2010, as temperatures have been warming and the warblers have been breeding earlier, they have developed larger and more attractive wing bars. However, by the time the breeding season is over, their wing bars have shrunk to be no larger than those of previous generations, because warmer summers increase feather wear. The opposing effects of a warmer spring and a warmer summer on wing-bar size mean that the Hume's warblers start off each breeding season with false advertising due to their large wing bars, which they are unable to maintain because of increased feather wear over the course of a longer

breeding season. Climate change has effectively made a sexually selected signal unreliable.

In other cases, climate change has caused species ranges to shift as a result of who is mating with whom. Carolina and black-capped chickadees can interbreed, and form a narrow hybrid zone in Pennsylvania, where the two species ranges overlap. This zone has been moving north at a rate of over 0.8 km (half a mile) a year, as average temperatures increase with climate change. A similar shift is taking place in Scotland and Denmark, where the hybrid zone between carrion crows and hooded crows is also moving north at a comparable rate.

Unlike most of their chicken relatives, ptarmigan are monogamous. However, this does not exempt them from sexual selection, and rock ptarmigan males retain their snowy white plumage for about three weeks longer than females. This delayed moult is costly to males, as they forgo the advantages of camouflage in order to gain a mate, and are more likely to be picked off by predators once the snow melts. As soon as a female starts incubation, which means she is no longer fertile, her white mate rubs himself in mud and dust while waiting for his flashy white feathers to moult. One can imagine that with climate change, an earlier snowmelt will not bode well for rock ptarmigan, particularly the males.

TRADE-OFFS AND ENVIRONMENTAL EXTREMES

Bar-headed geese migrate over the Tibetan Plateau at an altitude of almost 4.8 km (3 miles), yet they barely increase their heart rate, and do not need to train the way humans unaccustomed to high altitudes attempt to acclimatise before climbing tall mountains. This is because they have evolved a special haemoglobin molecule (this is what makes blood red) that binds oxygen more efficiently than the haemoglobin in most birds or mammals. Some species are already specialists at surviving in extreme climates, but this does not necessarily make them more resilient in the face of climate change.

Different populations have evolved different solutions to the low oxygen concentrations at high altitudes. Rather like the bar-headed geese, hummingbirds in the Andes have haemoglobin that binds oxygen more efficiently. Songbirds that live in the Himalayas year-round, such as the green-backed tit and grey-winged blackbird, deal with the chronic shortage of oxygen by having more haemoglobin molecules per cell.

By contrast, migrants, such as blue-fronted redstarts, which only spend a couple of months breeding high in the mountains before migrating closer to sea level, where there is more oxygen, respond much as a flatlander human would when visiting a ski resort. They temporarily make more red blood cells to compensate for the low levels of oxygen. This solution comes at a cost, because more cells make blood thicker, increasing the risk of blood clots, and making it much harder to pump the viscous blood around the body.

Specialising in life at high altitude rather than retaining the flexibility of a migrant also comes at a cost. With warming temperatures, high-altitude residents have nowhere to move but upslope – until they run out of mountain. Surveys in the mountains of New Guinea have found that over 70 per cent of bird species now live 40 metres (131 feet) higher than they did in 1985, when the biologist Jared Diamond first recorded them. Similarly, in the Peruvian Andes, half the ridge-top specialists have disappeared entirely, while nearly all the species living in the mountains are rarer than before. This is partly because mountains narrow as they ascend, so even if the birds shift their ranges upwards to cope with warming temperatures, they effectively shrink the area of their range.

Moving uphill with rising temperatures

Bird species in the Peruvian Andes have shifted their ranges up the mountains with warming temperatures. (1) The common scale-backed antbird has expanded its range by 17 per cent. (2) The versicoloured barbet has moved upslope, causing its range to shrink by 66 per cent. (3) The variable antshrike, a former mountaintop specialist, was not resighted in 2017.

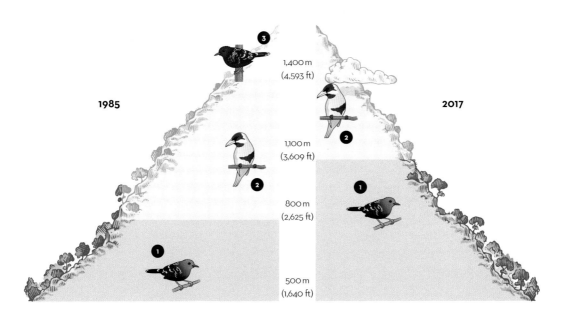

1985

2017

1,400 m
(4,593 ft)

1,100 m
(3,609 ft)

800 m
(2,625 ft)

500 m
(1,640 ft)

HABITAT SPECIALISTS

Sandgrouse are desert specialists distantly related to pigeons, and although their chicks are highly precocial, they 'milk' their father's specially adapted breast feathers for water. All but one sandgrouse species have evolved specialised feathers that soak up water like a sponge, so males can ferry water from water holes back to their chicks. When researchers took feathers from museum specimens and soaked them in water, they found that the male sandgrouse feathers held three times as much water as the feathers of other birds, and twice as much as female sandgrouse feathers. The Tibetan sandgrouse is the only species without these special feathers, because there appears to be an engineering trade-off between insulation and spongelike properties.

Another reproductive adaptation to different latitudes and habitats is egg pigment. Eggs exposed to more sunlight develop faster. Poultry scientists first discovered this in the 1960s, but since then, biologists have found that during daylight, the embryo in an egg has a metabolic rate one-and-a-half times faster than it does in darkness.

In addition, eggs with darker shells heat up more rapidly than brighter eggs that reflect more light. Across 634 species from most bird orders, species that live in colder climates have darker eggs, presumably to help keep them warm. This pattern only applies to species whose eggs were exposed to the sun, and not to those that lay their eggs in holes, burrows or enclosed nests. Rather surprisingly, temperature is a better predictor of how dark or bright eggs are than the need for camouflage in species encountering greater predation.

FLEXIBLE PARENTING

Shorebirds have some of the greatest variation in parental care among birds, both within and across species. Spotted sandpipers are the most widespread sandpiper breeding in North America. In this role-reversed species, females arrive at the breeding grounds before males, so as to tussle over territories. The females lay eggs for as many as three males, who then perform all other parenting duties. This leaves the females free to migrate south much earlier than their mates, who must wait for the chicks to grow more independent. Indeed, among shorebirds, species that only need one parent to successfully rear chicks tend to be polyandrous if the species also has long migratory journeys. Perhaps, after the massive investment in egg laying, female white-rumped sandpipers and other polyandrous females are unlikely to survive migration if they expend more energy in childcare as well. They simply leave the breeding grounds earlier, forcing the males to pick up the slack.

Most plover species share incubation duties, and males tend to take the night shift, possibly because they are sometimes more colourful and less well-camouflaged than females. A study looking at thirty-six plover species from across the world – from Canada to the Falkland Islands, and East Asia to southern Australia – found that when temperatures were more variable or higher, males performed a larger share of incubation, partly by taking on more of the daytime shift as well.

Opposite

Sandgrouse are desert specialists. They have feathers on their bellies adapted for soaking up water like sponges. This male Burchell's sandgrouse is just taking off from a waterhole in the Kalahari to ferry water to his chicks at the nest miles away.

MIGRATION DECISIONS

For most species, biologists have no idea which adaptations to changing climates are the result of flexible responses triggered by the environment, or more fixed behavioural rhythms encoded in the genes. Migration neatly encapsulates the notion that most adaptations to unpredictable environments involve both nature and nurture. Even animals with the most complicated genetic hardwiring for when and where to travel need environmental cues to calibrate these internal clocks.

The impulse to migrate must have evolved by natural selection from what biologists call *zugunruhe* (German for 'migratory restlessness'). The best guess for why migration is adaptive is that individuals making annual pilgrimages to capitalise on food bonanzas in temperate summers had more offspring than those that stayed put in the tropics all year round.

WHERE TO GO?

For some species, migration routes – as well as the impulse to migrate – are largely under genetic control. Common cuckoos seem to know instinctively where to migrate because they are raised by entirely different species, yet fly unguided to winter in Africa. Similarly, pectoral sandpipers leave their chicks before they fledge, but the fledglings are able to migrate unaided. In contrast, young cranes and goslings learn where to go on their first migratory flights regardless of whether their tutors are the same species, or a light aircraft operated by human foster parents showing them the way.

Blackcaps are Old World warblers that breed in northern Europe. Starting in the 1960s, birdwatchers in the British Isles began noticing increasing numbers of blackcaps in winter, when the birds were supposed to be in southern Europe. Biologists analysed the chemical composition of blackcap claws to pinpoint where the birds had been feeding before arriving back at their breeding grounds. This revealed that a subset of blackcaps breeding in central Europe had spent their winter growing fat on bird tables in the British Isles rather than travelling to their traditional wintering grounds in Spain or Portugal.

We now know that this new migratory direction and the shorter travel distance are the result of genetic differences, which are maintained by the mate choices the blackcaps make when back on their breeding grounds in Germany and Austria. Much as one can breed dogs to herd or to retrieve by selecting only those with the most extreme behaviours, biologists have also bred blackcaps that have less of an urge to migrate. Short-distance migrants such as blackcaps could have an advantage when it comes to evolving new migratory routes and distances.

WHEN TO GO?

In addition to internal hormonal clocks that are calibrated to daily rhythms, migratory birds have annual clocks that help them decide when to migrate.

Pied flycatchers have adapted to earlier springs through natural selection for a tweaking of their seasonal clock. Rather than relying purely on environmental cues such as temperature, flycatchers hatched in the wild but raised and kept in identical laboratory conditions bred nine days earlier in 2002 than those from the same population 21 years ago. Similarly, the 2002 males completed their winter flight moult and began developing gonads much earlier, which means that they migrated back and were ready to breed earlier than the 1981 birds.

This study suggests that a large component of the pied flycatcher annual clock is under genetic control, but can evolve rapidly under strong selection from a warming climate. However, these flycatchers are long-distance migrants, and cannot begin breeding earlier than when they arrive back from their wintering grounds in North Africa. Ultimately, the timing of migration could constrain how much these birds can shift their breeding clocks in response to climate change.

WHETHER TO GO?

Dark-eyed juncos in San Diego (US) have evolved to stop migrating from places that are sufficiently comfortable even in winter, making the bother of biannual travel unnecessary. Much like the different blackcap and pied flycatcher populations, these sedentary juncos are now genetically distinct from the migratory populations they originated from.

By contrast, some individual white storks that have traditionally migrated between Africa and Europe are now deciding to stay put because of the plethora of junk food in Portuguese landfills. As white storks are highly territorial during the breeding season, becoming a year-round resident at the nest also gives them a reproductive advantage in the spring. Not only does a nonmigrating stork save on the risks of long-distance travel, they also have a home advantage over migrants that arrive in spring to fight for coveted nest sites. Male storks that arrive back from wintering grounds earlier tend to bag the best nest sites and have the most offspring. There is no evidence that this behaviour is the result of natural selection on genetic variation for the tendency to migrate, as storks are long-lived, so many of these newly sedentary individuals would have been migrants in their youth.

Opposite

Some white storks in Portugal have stopped migrating south for the winter because they get enough food year round from landfills. This gives them first dibs on the most coveted nests in the spring.

MIGRATION MECHANICS

Most birds that breed at higher latitudes migrate to balmier places during the nonbreeding season. We still know shockingly little about how and why some species have evolved to pull off such feats of endurance. Arctic terns are the most extreme for the total length of their annual journeys, flying between the Arctic and Antarctic twice a year.

TRACKING BIRDS

Some tools that are used for tracking migrating birds include:

- Metal rings on bird legs
- GPS tags
- Satellite tags
- Geolocators
- VHF radio tags
- Micro data loggers

The most famous individual long-distance migrant is a female bar-tailed Godwit that wore a satellite transmitter and flew for eight days without pause between New Zealand and Alaska. We now know that all bar-tailed Godwits are capable of this feat of endurance. Even the diminutive ruby-throated hummingbird is capable of flying more than 2,000 km (1,243 miles) without pause.

In medieval times, people explained the seasonal disappearance of barnacle geese by assuming they turned into barnacles. Some of the earliest evidence in Europe that birds migrate long distances was the appearance, in 1822, of a white stork with an African hunting spear embedded in its neck. There are actually as many as twenty-five of these *pfeilstorch* (German for 'arrow stork') specimens documented. White storks are also the first migratory animals to be unintentionally tagged. For decades, biologists have relied on a similar principle by attaching small rings to bird legs in the hope of catching them again elsewhere. However, this only gives one or two locations at best, and the recapture rate is only about 1 per cent.

TRACKING TECHNOLOGY

Nowadays, biologists are able to track migrations with some truly impressive technology. Satellite tags allow us to follow the migrations of individual birds. Common cuckoo populations have been declining, but a combination of crowd-sourced funding and grants has allowed scientists in China to tag a handful of individuals. These birds were named by local schoolchildren, and followed with bated breath by people from all over the world as the signals from their solar-powered satellite tags showed where they were in real time. They even had their own Twitter feeds. The tags revealed that the individuals that breed in China fly through India to winter in Africa, connecting communities of people all along the migration route and beyond, such as in the UK, where this project originated.

Above

This white stork with an African hunting spear in its neck is one of several *pfeilstorch* (German for 'arrow stork') that provided some of the first evidence that birds migrated, rather than hibernating or metamorphosing, in the winter.

GEOLOCATORS

The downside of satellite transmitters is their size. By contrast, geolocators weigh a fraction of a gram and record a bird's position throughout migration, which helps to track smaller birds and to identify important stopover sites. However, like rings, these require the bird to be caught again before the data can be downloaded. An even lighter nanotag uses VHF radio waves and is small enough to fit on a dragonfly. Birds Canada is leading a collaboration called the Motus Wildlife Tagging System, with hundreds of receivers tracking nanotagged animals from the Arctic to South America.

Above

The study of migration has been revolutionised by tracking devices like this geolocator, light and small enough to fit on the backs of some of the smallest birds, such as this Connecticut warbler, which weighs about 14 grams (half an ounce).

DATA LOGGERS

Micro data loggers contain an accelerometer for measuring how fast a bird is moving, and a light sensor from which biologists can calculate location based on the time of the recording and the angle of the sun. These devices are so small that they weigh less than 1 gram (0.035 ounces), and have confirmed that common swifts spend ten months in the air without ever landing. They also revealed that the swifts, caught in Sweden, wintered in West Africa. By contrast, about 50,000 swifts had been ringed in Sweden over 100 years, and only one was ever recovered south of the Sahara.

By combining satellite trackers and light-sensitive data loggers with satellite images of greening foliage, a study in 2017 has shown that three species that migrate between the Palearctic and Africa – common cuckoos, red-backed shrikes and thrush nightingales – are surprisingly precise and plastic at tracking peaks in food over the course of their journeys. These birds were able to match their movements to fine-scale seasonal variations, both within their wintering grounds in Africa and as they migrated to Europe.

SYSTEM MAINTENANCE

Feathers are crucial equipment and must be moulted and maintained. Many migrants first fly to staging areas, where they moult into a new set of feathers. Moult can consume a quarter of the protein in a bird's body. A study of rufous-collared sparrows showed that individuals with higher levels of stress hormones circulating in their blood grew poorer quality feathers. In addition, birds can suppress their stress responses in order to grow good feathers, in spite of stressful environmental events such as the changing seasons.

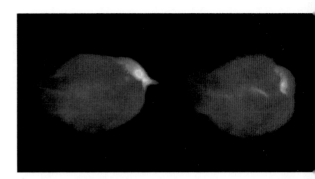

Although birds on migration are able to put half their brain to sleep at any one time, layover areas are crucial for many birds to recharge and refuel. As with everything in life, resting birds face trade-offs. By using thermal imaging cameras and measuring the respiratory rates of migrant warblers making a rest stop on Ponzu, an Italian island, biologists found that healthy, well-fed birds could afford to sleep with their heads exposed, and had a higher metabolic rate, which would allow them to react more quickly to danger. By contrast, their more bedraggled companions slept in a more energy-saving posture with their heads tucked in, which left them relatively vulnerable to predators. Using tracking technology to identify key places where migrants break a journey is crucial if we wish to save these species.

Above, right

The first thermal image shows a bird sleeping before it tucks its head into its feathers; the second reveals how once its head is tucked in, the heat loss is reduced, lowering its overall temperature.

Right

A yellow warbler asleep in the rocks.

FLIGHT FUEL

Migrants as tiny as hummingbirds, or as large as Godwits, double their weight before migrating. Crucial layover sites such as the Delaware Bay (US) allow red knots to refuel on horseshoe crab eggs, regaining half their initial weight in three weeks. Most of this is fat (the most efficient fuel for animals), and can be stored in massively enlarged livers.

Feeding frenzy

Red knots are among the migratory shorebirds that gather in vast numbers to feed on horseshoe crab eggs on Delaware Bay, on the east coast of America. Rising sea levels due to climate change will make this already declining shorebird even more vulnerable by removing some of their key refuelling sites.

Waterfowl, including many ducks and geese, naturally store this newly minted fat in their livers, creating a natural version of the French delicacy, *foie gras*. Hunters occasionally shoot pintail with livers comparable to farmed *foie gras*, after the ducks have gorged on rice in California in preparation for their autumn migration. A farmer in Spain has even found a way to harness this natural tendency for waterfowl to develop fatty livers, by providing a buffet of acorns and other favourite foods, and harvesting the livers of free-

ranging geese who come to gorge in preparation for a journey they will never make. This *foie gras* has won awards in blind taste tests against livers from birds force-fed in the traditional manner with a funnel.

The goose genome has extra copies of genes that rapidly make fat and store it in the liver, suggesting that these bird livers are specially adapted, seasonal fuel tanks. In addition, geese seem to suppress inflammation typically associated with steatosis (enlarged fatty liver) by having high blood concentrations of lactic acid produced by their gut bacteria. Understanding how waterfowl cope physiologically with massive fatty livers could help treat what is a disease in humans.

Also relevant to human health is how birds regulate the impulse to stuff themselves. Thrush nightingales use geomagnetic fields to trigger the impulse to overeat before their migration to eastern Africa. One of the hormones that controls appetite is ghrelin, and mammals and birds share this. By injecting ghrelin in wild garden warblers, biologists have shown that higher levels of this hormone decrease appetite and increase migratory restlessness. They also found that birds with naturally higher levels of the hormone had accumulated more fat stores.

NAVIGATION

Migrating birds use compass information from the sun, the stars and by sensing the earth's magnetic field. They also use landmarks, and homing pigeons use smells to know which direction to fly in when taken to an entirely new place. Most birds migrate in flocks, so individuals can also get more information in a group.

Below

During their migration south, a flock of sandhill cranes fly past the snow-covered peak of Denali, in Alaska.

Navigating collectively can occur in various ways. Just as a crowd can more accurately converge on the number of beans in a jar than any single individual's guess, pooled estimates from multiple animals is most likely to lead to the optimal migratory

route. This law of averages applies to migrating wild skylarks and common scoters (a type of duck), and has been demonstrated extensively with experiments on homing pigeons.

In smaller groups, informed individuals can lead naive ones on the right flight path. Taken a step further, this can result in the naive individuals learning from more experienced ones. Both these mechanisms apply to whooping cranes. The most mind-boggling version of collective navigation occurs via emergent sensing, where no individual in the flock can single-handedly detect an environmental gradient, but collectively, the flock can do so, with each bird acting as a sensor within a large network.

White storks are known to use a combination of all these navigational mechanisms. By fitting twenty-seven juveniles with GPS trackers and accelerometers, scientists from a Max Planck institute followed not just where each member of the flock went, but also its speed. They found that some of the storks were more likely to lead the way to thermals, which allowed them to flap less and conserve energy. By contrast, other storks benefited from following these leaders. Curiously, the follower storks also tended to lag or drop off, leading to much more high-energy flapping, and eventually migrated shorter distances than the leaders. The follower storks tended to end up in southern Europe, whereas the more efficient leaders flew all the way to Africa.

LIGHT POLLUTION

Unfortunately, migrating birds are unaccountably drawn to and disoriented by bright lights. We know this because lighthouses have attracted flocks of migrating songbirds for centuries. In addition, the lights within tall buildings attract migrating flocks that collide with the glass. It can take less than a week for about half the birds migrating through North America to pass through a single city. Volunteers have been counting dead birds at the base of buildings since the 1990s in cities such as New York, and an estimated 600 million die every year from flying into skyscrapers. Biologists have estimated that artificial light at night has been increasing by 5–10 per cent every year in most of North America and Europe. One solution that has helped greatly is the use of glass with tiny patterns visible to birds, which has cut collisions by 90 per cent.

BIG BIRD DATA

Birds are uniquely positioned to help us understand and cope with the impact of a global climate crisis. Their vulnerability makes them a good environmental early-warning system. Vast spaces linked by the long migratory journeys of many species help us to understand the global scale of climate change.

Birds' visibility and charisma make them part of the solution. For no other group of living things do scientists have reams of recordings that stretch back for decades, almost all faithfully gathered by volunteers worldwide. In addition to revealing dramatic and depressing declines in bird numbers, this data presents the chance to use technology to right some wrongs.

Most birds migrate at night, so biologists have started using existing weather radar technology to monitor large flocks. Some of the latest dual-polarization radar sends out two microwave beams instead of just one, allowing meteorologists to distinguish between rain, hail and snow within a storm. Ornithologists are using the same technology to see details such as which way a bird's bill is pointed.

Another radar technique called NEXRAD catches birds as they take off for their migratory flights, which allows biologists to identify important patches of habitat that could be crucial rest stops. Doppler radar data allows scientists to reconstruct past migrations. Their findings are rather discouraging, documenting drops of 4 per cent in migrant bird biomass every year.

Now, with citizen science platforms such as eBird, scientists can combine a multitude of recordings with information from radar, which shows how flocks move. This allows groups like the Cornell Lab of Ornithology to produce maps that animate the annual migrations of different bird species across continents. BirdCast uses radar scans and eBird submissions to predict bird movements, just like a weather forecast, but for bird migrations. Scientists are also combining this information with satellite images showing light pollution levels, in order to forecast the areas most likely to disorient migratory birds.

Conservationists are also working with governments to use this data to predict key flight paths, thereby minimising migratory massacres from hazards such as light pollution or aeroplanes. Bird avoidance systems from many countries including Israel, Poland and Germany, have collectively reduced collisions between birds and military aircraft by 45 per cent.

eBird data has revealed that songbirds migrate very differently from the larger and more classically studied migrants, such as shorebirds and waterfowl. Rather than going back and forth along the same route, these relatively diminutive travelers are more flexible. They also migrate in a loop, taking a different route on their journeys to and from breeding grounds, so as to take full advantage of tailwinds and to minimise headwinds.

Using computers to translate a variety of radar data, biologists from Cornell showed that shorter-distance migrants that winter in the lower forty-eight states of America and breed north of the US-Canada border were less likely to return to breed than longer-distance migrants that winter in South and Central America. This result is surprising, because one would expect shorter journeys to contain fewer hazards.

Above

Baltimore orioles festoon a tree during a 'fallout' from a storm, while they break their journey during their migration north in the spring.

UNDERSTANDING ANIMAL MOVEMENTS

Machine learning provides the next big step in understanding animal movements. By first training artificial intelligence programs on reams of bird recordings, much as one would train a human brain to distinguish birdsong from background noise, researchers are able to have computers translate sound recordings from bird-breeding grounds in the Arctic directly into arrival dates. This allows conservationists to know how migrant birds are responding to climate change without having to personally record or decipher anything. At present, the algorithm can distinguish birds from other noises, such as the wind or machines. The next step is to train the computer algorithms to identify individual bird species.

Similar methods could be used to identify and predict migratory rest stops, breeding grounds and the timing of bird movements, to inform governments and conservation groups about which places to concentrate on saving. Helping birds cope with climate is a global problem, not just because climate change is a global phenomenon, but also because migratory birds link multiple nations by travelling through them. Just one weak link is enough to precipitate a systemic collapse.

SANDPIPERS AT RISK

There are fewer than 500 spoon-billed sandpipers left in the world.

Some conservation efforts are proving successful. The rare and elusive Bicknell's thrush, which only breeds on mountaintops in New England, is now being protected at one of its key wintering grounds in the Dominican Republic. A massive outreach effort in northern India has turned what used to be a killing ground for Amur falcons into a sanctuary, allowing these charismatic little falcons to continue the longest-known migration of any raptor in the world. I have seen these beautiful birds roosting in Zambia, and they fly all the way from breeding grounds in China and Siberia. The key protected rest stop is in Nagaland, where the falcons refuel on trillions of seasonal termites before flying 3,900 km (2,400 miles) across the Indian Ocean. The hope is that ecotourists will now flock to see the migrating falcons, rewarding the people of Nagaland for giving up a traditional mass hunt of the birds.

Most heartening is that the Chinese government has agreed to halt land reclamation in the Yellow Sea, which is a key rest stop for birds on the East Asian–Australasian flyway. Shorebirds from Siberia and Alaska migrate down this flyway to winter in Australia, and the mudflats along the coast of the Yellow Sea have provided a crucial place for migrating shorebirds to refuel. I have witnessed the impact of land reclamation on these mudflats over the last 20 years, as migrant shorebird numbers dwindle in the wetland nature reserve that I visit annually on my own migration back to Singapore. Habitat loss on the Yellow Sea coast has caused Far Eastern curlew numbers to plummet by 80 per cent.

Birdwatchers are essential contributors to conservation efforts. Their love of birds mean they cluster to see rare birds that get lost on migration, and keep tabs on when and where they ever saw a bird. I am hopeful that we can continue to contribute collectively to accumulating big data that will help to save the birds we enjoy.

Above

Land reclamation and offshore wind turbines pose serious obstacles to migratory shorebirds, such as these dunlin, who rely on traditional staging grounds to refuel during their migrations.

BIBLIOGRAPHY

BOOKS

Black, J. M. (1996). *Partnerships in Birds: the Study of Monogamy.* Oxford University Press.

Davies, N. B. (2000). *Cuckoos, Cowbirds and Other Cheats.* T. & A. D. Poyser.

Davies, N. B. (1992). *Dunnock Behaviour and Social Evolution.* Oxford University Press.

Davies, N. B., Krebs, J. R., West, S. A. (2012). *An Introduction to Behavioural Ecology.* 4th edition, Wiley-Blackwell.

Koenig, W. D., Dickinson, J. L. (Ed.). (2016). *Cooperative Breeding in Vertebrates: Studies of Ecology, Evolution, and Behavior.* Cambridge University Press.

Payne, R. B., Sorenson, M. D. (2005). *The Cuckoos.* Oxford University Press.

JOURNAL ARTICLES

Abbey-Lee, R. N., Dingemanse, N. J. (2019). Adaptive individual variation in phenological responses to perceived predation levels. *Nature Communications, 10,* 5667.

Aplin, L. M., et al. (2017). Conformity does not perpetuate suboptimal traditions in a wild population of songbirds. *PNAS, 114,* 7830–7837.

Aplin, L. M., et al. (2015). Experimentally induced innovations lead to persistent culture via conformity in wild birds. *Nature, 518,* 538–541.

Araya-Salas, M., et al. (2018). Spatial memory is as important as weapon and body size for territorial ownership in a lekking hummingbird. *Scientific Reports, 8,* 2001.

Ashton, B. J., et al. (2018). Cognitive performance is linked to group size and affects fitness in Australian magpies. *Nature, 554,* 364–367.

Baldwin, M. W., et al. (2014). Evolution of sweet taste perception in hummingbirds by transformation of the ancestral umami receptor. *Science, 345,* 929–933.

Bearhop, S., et al. (2005). Assortative mating as a mechanism for rapid evolution of a migratory divide. *Science, 310,* 502–504.

Becciu, P., et al. (2019). Environmental effects on flying migrants revealed by radar. *Ecography, 42,* 942–955.

Bell, B. A., et al. (2018). Influence of early-life nutritional stress on songbird memory formation. *Proceedings of the Royal Society B, 285,* 20181270.

Boeckle, M., Clayton, N. S. (2017). A raven's memories are for the future. *Science, 357,* 126–127.

Bosse, M., et al. (2017). Recent natural selection causes adaptive evolution of an avian polygenic trait. *Science, 358,* 365–368.

Both, C., Visser, M. E. (2001). Adjustment to climate change is constrained by arrival date in a long-distance migrant bird. *Nature, 411,* 296–298.

Brown, C. R., Bomberger Brown, M. (2013). Where has all the road kill gone? *Current Biology, 23,* 233–234.

Bryan, R. D., Wunder, M. B. (2014). Western burrowing owls (*Athene cunicularia hypugaea*) eavesdrop on alarm calls of black-tailed prairie dogs (*Cynomys ludovicianus*). *Ethology, 120,* 180–188.

Bulla, Martin, et al. (2016). Unexpected diversity in socially synchronized rhythms of shorebirds. *Nature, 540,* 109–113.

Burley, N. T., Symanski, R. (1998). "A taste for the beautiful": Latent aesthetic mate preferences for white crests in two species of Australian grassfinches. *American Naturalist 15,* 792–802.

Campagna, L., et al. (2017). Repeated divergent selection on pigmentation genes in a rapid finch radiation. *Science Advances, 3,* e1602404.

Carleial, R., et al. (2020). Dynamic phenotypic correlates of social status and mating effort in male and female red junglefowl, *Gallus gallus. Journal of Evolutionary Biology, 33,* 22–40.

Carlo, T. A., Tewksbury, J. J. (2014). Directness and tempo of avian seed dispersal increases emergence of wild chiltepins in desert grasslands. *Journal of Ecology, 102,* 248–255.

Carpenter, J. K., et al. (2019). Long seed dispersal distances by an inquisitive flightless rail (*Gallirallus australis*) are reduced by interaction with humans. *Royal Society Open Science, 6,* 190397.

Child, M. F., et al. (2012). Investigating a link between bill morphology, foraging ecology

and kleptoparasitic behaviour in the fork-tailed drongo. *Animal Behaviour, 84,* 1013–1022.

Clayton, N. S. & Emery, N. J. (2007). The social life of corvids. *Current Biology, 17,* 652–656.

Cornwallis, C. K., O'Connor, E. A. (2009). Sperm: Seminal fluid interactions and the adjustment of sperm quality in relation to female attractiveness. *Proceedings of the Royal Society B, 276,* 3467–3475.

Couzin, I. D. (2018). Collective animal migration. *Current Biology, 28,* 952–1008.

Da Silva, A., Kempenaers, B. (2017). Singing from north to south: Latitudinal variation in timing of dawn singing under natural and artificial light conditions. *Journal of Animal Ecology, 86,* 1286–1297.

Davis, H. T., et al. (2018). Nest defense behavior of Greater Roadrunners (*Geococcyx californianus*) in south Texas. *Wilson Journal of Ornithology, 130,* 788–792.

Depeursinge, A., et al. (2019). The multilevel society of a small-brained bird. *Current Biology, 29,* 1120–1121.

Drummond, H., et al. (2016). An unsuspected cost of mate familiarity: increased loss of paternity. *Animal Behaviour, 111,* 213–216.

Earp, S. E. & Maney, D. L. (2012). Birdsong: Is it music to their ears? *Frontiers in Evolutionary Neuroscience, 4,* 14.

Echeverría, V., et al. (2018). Pre-basic molt, feather quality, and modulation of the adrenocortical response to stress in two populations of rufous-collared sparrows *Zonotrichia capensis. Journal of Avian Biology,* e01892.

Evans, S. R. & Gustafsson, L. (2017). Climate change upends selection on ornamentation in a wild bird. *Nature Ecology & Evolution, 1,* 1–5.

Fallow, P. M., Magrath, R. D. (2010). Eavesdropping on other species: mutual interspecific understanding of urgency information in avian alarm calls. *Animal Behaviour, 79,* 411–417.

Farine, D. R., et al. (2019). Early-life social environment predicts social network position in wild zebra finches. *Proceedings of the Royal Society B, 286,* 20182579.

Fehér, O., et al. (2009). De novo

establishment of wild-type song culture in the zebra finch. *Nature, 459*, 564–568.

Felice, R. N., O'Connor, P. M. (2014). Ecology and caudal skeletal morphology in birds: The convergent evolution of pygostyle shape in underwater foraging taxa. *PLoS One, 9*(2).

Felice, R. N., et al. (2019). Dietary niche and the evolution of cranial morphology in birds. *Proceedings of the Royal Society B, 286,* 20182677.

Ferretti, A., et al. (2019). Sleeping unsafely tucked in to conserve energy in a nocturnal migratory songbird. *Current Biology, 29,* 2766–2772.

Firth, J. A., et al. (2015). Experimental evidence that social relationships determine individual foraging behavior. *Current Biology, 25,* 3138–3143.

Firth, J. A., et al. (2018). Personality shapes pair bonding in a wild bird social system. *Nature Ecology & Evolution, 2,* 1696–1699.

Flack, A., et al. (2018). From local collective behavior to global migratory patterns in white storks. *Science, 360,* 911–914.

Found, R. (2017). Interactions between cleaner-birds and ungulates are personality dependent. *Biology Letters, 13,* 20170536.

Francis, M. L., et al. (2018). Effects of supplementary feeding on interspecific dominance hierarchies in garden birds. *PLoS One, 13*(9).

Gahr, M. Vocal communication: Decoding sexy songs. (2018). *Current Biology, 28,* 306–327.

Galván, I., et al. (2019). Unique evolution of vitamin A as an external pigment in tropical starlings. *Journal of Experimental Biology, 222,* 205229.

Gautier, P., et al. (2008). The presence of females modulates the expression of a carotenoid-based sexual signal. *Behavioral Ecology and Sociobiology, 62,* 1159–1166.

George, J. M., et al. (2019). Acute social isolation alters neurogenomic state in songbird forebrain. *PNAS, 22,* 201820841.

Gibson, R. M., et al. (1991). Mate choice in lekking sage grouse revisited: The roles of vocal display, female site fidelity, and copying. *Behavioral Ecology, 2,* 165–180.

Gilbert, N. I., et al. (2015). Are white storks addicted to junk food? Impacts of landfill use on the movement and behaviour of resident white storks (*Ciconia ciconia*) from a partially migratory population. *Movement Ecology, 4,* 7.

Goodale, E., Kotagama, S. W. (2008). Response to conspecific and heterospecific alarm calls in mixed-species bird flocks of a Sri Lankan rainforest. *Behavioral Ecology, 19,* 887–894.

Goymann, W., et al. (2017). Ghrelin affects stopover decisions and food intake in a long-distance migrant. *PNAS, 114,* 1946–1951.

Grant, B. R., Grant, P. R. (2010). Songs of Darwin's finches diverge when a new species enters the community. *PNAS, 107,* 20156–20163.

Grant, P. R., Grant, B. R. (2008). Pedigrees, assortative mating and speciation in Darwin's finches. *Proceedings of the Royal Society B, 275,* 661–668.

Greeney, H. F., et al. (2015). Trait-mediated trophic cascade creates enemy-free space for nesting hummingbirds. *Science Advances, 1,* e1500310.

Gwinner, H., et al. (2018). 'Green incubation': Avian offspring benefit from aromatic nest herbs through improved parental incubation behaviour. *Proceedings of the Royal Society B, 285,* 20180376.

Hedenström, A., et al. (2016). Annual 10-Month Aerial Life Phase in the Common Swift *Apus apus*. *Current Biology, 26,* 3066–3070.

Heinsohn, R., et al. (2017). Tool-assisted rhythmic drumming in palm cockatoos shares key elements of human instrumental music. *Science Advances, 3,* e1602399.

Helm, B., et al. (2019). Evolutionary response to climate change in migratory pied flycatchers. *Current Biology, 29,* 3714–3719.

Hijglund, J., et al. (1995). Mate-choice copying in black grouse. *Animal Behaviour, 49,* 1627–1633.

Hill, G. E., McGraw, K. J. (2004). Correlated changes in male plumage coloration and female mate choice in cardueline finches. *Animal Behaviour, 67,* 27–35.

Hinks, A. E., et al. (2015). Scale-dependent phenological synchrony between songbirds and their caterpillar food source. *American Naturalist, 186,* 85–97.

Hoffmann, S., et al. (2019). Duets recorded in the wild reveal that interindividually coordinated motor control enables cooperative behavior. *Nature Communications, 10,* 2577.

Igic, B., et al. (2015). Crying wolf to a predator: Deceptive vocal mimicry by a bird protecting young. *Proceedings of the Royal Society B, 282,* 20150798.

Ihle, M., et al. (2015) Fitness benefits of mate choice for compatibility in a socially monogamous species. *PLoS Biology, 13,* 1002248.

Jelbert, S. A., et al. (2018). Mental template matching is a potential cultural transmission mechanism for New Caledonian crow tool manufacturing traditions. *Scientific Reports, 8,* 1–8.

Jetz, W., et al. (2014). Global distribution and conservation of evolutionary distinctness in birds. *Current Biology, 24,* 919–930.

Joanne, R., et al. (2019). Spontaneity and diversity of movement to music are not uniquely human. *Current Biology, 29,* 603–622.

Kareklas, K., et al. (2019). Signal complexity communicates aggressive intent during contests, but the process is disrupted by noise. *Biology Letters, 15,* 20180841.

Karubian, J., et al. (2011). Bill coloration, a flexible signal in a tropical passerine bird, is regulated by social environment and androgens. *Animal Behaviour, 81,* 795–800

Keagy, J., et al. (2009). Male satin bowerbird problem-solving ability predicts mating success. *Animal Behaviour, 78,* 809–817.

Kempenaers, B., et al. (2018). Interference competition pressure predicts the number of avian predators that shifted their timing of activity. *Proceedings of the Royal Society B, 285,* 20180744.

Kiss, D., et al. (2013). The relationship between maternal ornamentation and feeding rate is explained by intrinsic nestling quality. *Behavioral Ecology and Sociobiology, 67,* 185–192.

Kniel, N., et al. (2015). Novel mate preference through mate-choice copying in zebra finches: Sexes differ. *Behavioral Ecology, 26,* 647–655.

Knight, K. (2016). Mystery of broadbills' wing song revealed. *Journal of Experimental Biology, 219,* 905.

La Sorte, F. A., et al. (2018). Seasonal associations with novel climates for North American migratory bird populations. *Ecology Letters, 21,* 845–856.

Lamichhaney, S., et al. (2018). Rapid hybrid speciation in Darwin's finches. *Science, 359,* 224–228.

Lanctot, R. B., et al. (1998). Male traits, mating tactics and reproductive success in the buff-breasted sandpiper, *Tryngites subruficollis*. *Animal Behaviour, 56,* 419–432.

Legg, E. W., Clayton, N. S. (2014). Eurasian jays (*Garrulus glandarius*) conceal caches from onlookers. *Animal. Cognition, 17,* 1223–1226.

Leniowski, K., Wegrzyn, E. (2018). Synchronisation of parental behaviours reduces the risk of nest predation in a socially monogamous passerine bird. *Scientific Reports, 8,* 1–9.

Liévin-Bazin, A., et al. (2019). Food sharing and affiliation: An experimental and longitudinal study in cockatiels (*Nymphicus hollandicus*). *Ethology, 125,* 276–288.

Lilly, M. V., et al. (2019). Eavesdropping grey squirrels infer safety from bird chatter. *PLoS One, 14,* e0221279.

Ling, H., et al. (2019). Costs and benefits of social relationships in the collective motion of bird flocks. *Nature Ecology & Evolution, 3,* 943–948.

Ling, H., et al. (2019). Behavioural plasticity and the transition to order in jackdaw flocks. *Nature Communications, 10,* e5174.

Lu, L., et al. (2015). The goose genome sequence leads to insights into the evolution of waterfowl and susceptibility to fatty liver. *Genome Biology, 16,* 89.

Winterbottom M., et al. (2001). The phalloid organ, orgasm and sperm competition in a polygynandrous bird: The red-billed buffalo weaver (*Bubalornis niger*). *Behavioral Ecology and Sociobiology, 50,* 474–482.

Macarthur, R. H. (1958). Population ecology of some warblers of northeastern coniferous forests. *Ecology, 39,* 599–619.

Madden, J. (2001). Sex, bowers and brains. *Proceedings of the Royal Society B, 268,* 833–838.

Magrath, R. D., Bennett, T. H. (2012). A micro-geography of fear: Learning to eavesdrop on alarm calls of neighbouring heterospecifics. *Proceedings of the Royal Society B, 279,* 902–909.

Maia, R., et al. (2013). Key ornamental innovations facilitate diversification in an avian radiation. *PNAS, 110,* 10687–92.

Malpass, J. S., et al. (2017). Species-dependent effects of bird feeders on nest predators and nest survival of urban American robins and northern cardinals. *Condor 119,* 1–16.

Mariette, M. M., Buchanan, K. L. (2016). Prenatal acoustic communication programs offspring for high posthatching temperatures in a songbird. *Science, 353,* 812–814.

Mathot, K. J., et al. (2017). Provisioning tactics of great tits (*Parus major*) in response to long-term brood size manipulations differ across years. *Behavioral Ecology, 28,* 1402–1413.

Mcqueen, A., et al. (2019). Evolutionary drivers of seasonal plumage colours: colour change by moult correlates with sexual selection, predation risk and seasonality across passerines. *Ecology Letters, 22,* 1838–1849.

Mets, D. G., Brainard, M. S. (2019). Learning is enhanced by tailoring instruction to individual genetic differences. *eLife, 8,* e47216.

Miller, E. T., et al. (2017). Fighting over food unites the birds of North America in a continental dominance hierarchy. *Behavioral Ecology, 28,* 1454–1463.

Mocha, Y., et al. (2018). Why hide? Concealed sex in dominant Arabian babblers (*Turdoides squamiceps*) in the wild. *Evolution and Human Behavior, 39,* 575–582.

Mocha, Y. Ben, Pika, S. (2019). Intentional presentation of objects in cooperatively breeding Arabian babblers (*Turdoides squamiceps*). *Frontiers in Ecology and Evolution, 7,* 87.

Moks, K., et al. (2016). Predator encounters have spatially extensive impacts on parental behaviour in a breeding bird community. *Proceedings of the Royal Society B, 283,* 20160020.

Morales, J., et al. (2018). Maternal programming of offspring antipredator behavior in a seabird. *Behavioral Ecology, 29,* 479–485.

Mutzel, A., et al. (2019). Effects of manipulated levels of predation threat on parental provisioning and nestling begging. *Behavioral Ecology, 30,* 1123–1135.

Navalón, G., et al. (2020). The consequences of craniofacial integration for the adaptive radiations of Darwin's finches and Hawaiian honeycreepers. *Nature Ecology & Evolution, 4,* 270–278.

Nelson-Flower, M. J., Ridley, A. R. (2016). Nepotism and subordinate tenure in a cooperative breeder. *Biology Letters, 12,* 20160365.

Nilsson, C., et al. (2019). Revealing patterns of nocturnal migration using the European weather radar network. *Ecography, 42,* 876–886.

Noguera, J. C., Velando, A. (2019). Bird embryos perceive vibratory cues of predation risk from clutch mates. *Nature Ecology & Evolution, 3,* 1225–1232.

Nunn, C. L., et al. (2011). Mutualism or parasitism? Using a phylogenetic approach to characterize the oxpecker-ungulate relationship. *Evolution, 65,* 1297–1304.

Oliver, R. Y., et al. (2018). Eavesdropping on the Arctic: Automated bioacoustics reveal dynamics in songbird breeding phenology. *Science Advances, 4,* 1084.

Patricelli, G. L., Krakauer, A. H. (2010). Tactical allocation of effort among multiple signals in sage grouse: An experiment with a robotic female. *Behavioral Ecology, 21,* 97–106.

Pérez-Camacho, L., et al. (2018). Structural complexity of hunting habitat and territoriality increase the reversed sexual size dimorphism in diurnal raptors. *Journal of Avian Biology, 49,* e10745.

Perrot, C., et al. (2016). Sexual display complexity varies non-linearly with age and predicts breeding status in greater flamingos. *Scientific Reports, 6,* 36242.

Phillips, R. A., et al. (2004). Seasonal sexual segregation in two Thalassarche albatross species: Competitive exclusion, reproductive role specializaion or foraging niche divergence? *Proceedings of the Royal Society B, 271,* 1283–1291.

Piersma, T. (1998). Phenotypic flexibility during migration: Optimization of organ size contingent on the risks and rewards of fueling and flight? *Journal of Avian Biology, 29,* 511.

Pika, S., Bugnyar, T. (2011). The use of referential gestures in ravens (*Corvus corax*) in the wild. *Nature Communications, 2,* 560.

Plantan, T., et al. (2013). Feeding preferences of the red-billed oxpecker, *Buphagus erythrorhynchus*: A parasitic mutualist? *African Journal of Ecology, 51,* 325–336.

Plummer, K. E., et al. (2019). The composition of British bird communities is associated with long-term garden bird feeding. *Nature Communications, 10,* 1–8.

Potvin, D. A., Clegg, S. M. (2015). The relative roles of cultural drift and acoustic adaptation in shaping syllable repertoires of island bird populations change with time since colonization. *Evolution, 69,* 368–380.

Potvin, D. A., et al. (2018). Birds learn socially to recognize heterospecific alarm calls by acoustic association. *Current Biology, 28,* 2632–2637.

Price, T. D., et al. (2014). Niche filling slows the diversification of Himalayan songbirds. *Nature*, 509, 222–225.

Price, T. D., et al. (2000). The imprint of history on communities of North American and Asian warblers. *American Naturalist*, 156, 354–367.

Prokop, Z. M., et al. (2012). Meta-analysis suggests choosy females get sexy sons more than 'good genes'. *Evolution*, 66, 2665–2673.

Pulido, F., Berthold, P. (2010). Current selection for lower migratory activity will drive the evolution of residency in a migratory bird population. *PNAS*, 107, 7341–7346.

Qvarnström, A., et al. (2004). Female collared flycatchers learn to prefer males with an artificial novel ornament. *Behavioral Ecology*, 15, 543–548.

Radford, A. N., Du Plessis, M. A. (2003). Bill dimorphism and foraging niche partitioning in the green woodhoopoe. *Journal of Animal Ecology*, 72, 258–269.

Rotics, S., et al. (2018). Early arrival at breeding grounds: Causes, costs and a trade-off with overwintering latitude. *Journal of Animal Ecology*, 87, 1627.

Rutz, C., et al. (2010). The ecological significance of tool use in new Caledonian crows. *Science*, 329, 1523–1526.

Rutz, C., et al. (2016). Discovery of species-wide tool use in the Hawaiian crow. *Nature*, 537, 403–407.

San-Jose, L. M., et al. (2019). Differential fitness effects of moonlight on plumage colour morphs in barn owls. *Nature Ecology & Evolution*, 3, 1331–1340.

Sánchez-Macouzet, O., et al. (2014). Better stay together: Pair bond duration increases individual fitness independent of age-related variation. *Proceedings of the Royal Society B*, 281, 20132843.

Sánchez-Macouzet, O., Drummond, H. (2011). Sibling bullying during infancy does not make wimpy adults. *Biology Letters*, 7, 869–871.

Santangeli, A., et al. (2017). Stronger response of farmland birds than farmers to climate change leads to the emergence of an ecological trap. *Biological Conservation*, 17, 166–172.

Savoca, M. S., et al. (2016). Marine plastic debris emits a keystone infochemical for olfactory foraging seabirds. *Science Advances*, 2, e1600395.

Smith, S. H., et al. (2017). Earlier nesting by generalist predatory bird is associated with human responses to climate change. *Journal of Animal Ecology*, 86, 98–107.

Snyder, K. T., Creanza, N. (2019). Polygyny is linked to accelerated birdsong evolution but not to larger song repertoires. *Nature Communications*, 10, 884.

Song, S. J., et al. (2019). Is there convergence of gut microbes in blood-feeding vertebrates? *Philosophical Transactions of the Royal Society B*, 374, 20180249.

Spottiswoode, C. N., et al. (2016). Reciprocal signaling in honeyguide-human mutualism. *Science*, 353, 387–389.

Suetsugu, K., et al. (2015). Avian seed dispersal in a mycoheterotrophic orchid *Cyrtosia septentrionalis*. *Nature Plants*, 1, 1–2.

Suzuki, T. N. (2018). Alarm calls evoke a visual search image of a predator in birds. *PNAS*, 115, 1541–1545.

Suzuki, T. N., et al. (2017). Wild birds use an ordering rule to decode novel call sequences. *Current Biology*, 27, 2331–2336.e3.

Tanaka, K. D., et al. (2005). Yellow wing-patch of a nestling Horsfield's hawk cuckoo *Cuculus fugax* induces miscognition by hosts: Mimicking a gape? *Journal of Avian Biology*, 36, 461–464.

Taylor, S. A., et al. (2014). Climate-mediated movement of an avian hybrid zone. *Current Biology*, 24, 671–676.

Tebbich, S., et al. (2001). Do woodpecker finches acquire tool-use by social learning? *Proceedings of the Royal Society B*, 268, 2189–2193.

Templeton, C. N., Greene, E. (2007). Nuthatches eavesdrop on variations in heterospecific chickadee mobbing alarm calls. *PNAS*, 104, 5479–82.

Templeton, C., et al. (2005). Allometry of alarm calls: Black-capped chickadees encode information about predator size. *Science*, 308, 1934–1937.

Thomas, D. B., et al. (2014). Ancient origins and multiple appearances of carotenoid-pigmented feathers in birds. *Proceedings of the Royal Society B*, 281, 20140806.

Thorup, K., et al. (2017). Resource tracking within and across continents in long-distance bird migrants. *Science Advances*, 3, e1601360.

Van de Pol, M., et al. (2009). Fluctuating selection and the maintenance of individual and sex-specific diet specialization in free-living oystercatchers. *Evolution*, 64, 836–851.

Van Doren, B. M., Horton, K. G. (2018). A continental system for forecasting bird migration. *Science*, 361, 1115–1118.

van Gasteren, H., et al. (2019). Aeroecology meets aviation safety: Early warning systems in Europe and the Middle East prevent collisions between birds and aircraft. *Ecography*, 42, 899–911.

Van Gils, J. A., et al. (2016). Body shrinkage due to Arctic warming reduces red knot fitness in tropical wintering range. *Science*, 352, 819–821.

van Lawick-Goodall, J., van Lawick-Goodall, H. (1966). Use of tools by the Egyptian vulture, *Neophron percnopterus*. *Nature*, 212, 1468–1469.

Vergara, P., et al. (2012). The condition dependence of a secondary sexual trait is stronger under high parasite infection level. *Behavioral Ecology*, 23, 502–511.

Wiemann, J., et al. (2018). Dinosaur egg colour had a single evolutionary origin. *Nature*, 563, 555–558.

Wiley, E. M., Ridley, A. R. (2018). The benefits of pair bond tenure in the cooperatively breeding pied babbler (*Turdoides bicolor*). *Ecology and Evolution*, 8, 7178–7185.

Wisocki, P. A., et al. (2020). The global distribution of avian eggshell colours suggest a thermoregulatory benefit of darker pigmentation. *Nature Ecology & Evolution*, 4, 148–155.

Yosef, R., et al. (2011). Set a thief to catch a thief: Brown-necked raven (*Corvus ruficollis*) cooperatively kleptoparasitize Egyptian vulture (*Neophron percnopterus*). *Naturwissenschaften*, 98, 443–446.

Yu, J., et al. (2019). Heterospecific alarm-call recognition in two warbler hosts of common cuckoos. *Animal Cognition*, 22, 1149–1157.

Zenzal, T. J., Moore, F. R. (2016). Stopover biology of ruby-throated hummingbirds (*Archilochus colubris*) during autumn migration. *The Auk*, 133, 237–250.

Zub, K., et al. (2017). Silence is not golden: The hissing calls of tits affect the behaviour of a nest predator. *Behavioral Ecology and Sociobiology*, 71, 79.

Zuk, M., Johnsen, T. S. (2000). Social environment and immunity in male red jungle fowl. *Behavioral Ecology*, 11, 146–153.

INDEX

ACKNOWLEDGEMENTS

This book owes its existence to many more than can be thanked here. UniPress, especially Kate Shanahan, Nigel Browning and Kate Duffy, have made the entire project possible. Special thanks to Ben Sheldon and Doug Mock for insightful comments. For staunch support and stimulating conversation, thanks to Bill Burnside, Nancy Curtis, Doug Emlen, David Haig, C. J. Huang, Dino Martins, Aileen Nielsen, Celeste Peterson, Naomi Pierce, Margaret Leng Tan, Laura Taylor, the 'Grey Jays', and more friends and family than I can name. Most heartfelt thanks to Ana and the birds for every inspiring and joyous daily walk.

UniPress Books would like to thank Wayne Blades for the elegant design and Kate Osborne for her beautiful watercolours.

PICTURE CREDITS

Thank you to the following sources for providing the images featured in this book.

Alamy/AGAMI Photo Agency: 47; All Canada Photos: 192; Arco Images GmbH: 151; Blickwinkel: 90; Danita Delimont 207B; Mick Durham FRPS: 172; Lisa Geoghegan: 163; imageBROKER: 106, 144L; Don Johnston BI: 109; Juniors Bildarchiv GmbH: 144R; Francois Loubser: 198; McDonald Wildlife Photography: 168; Minden Pictures: 104, 195; National Geographic Image Collection: 73R; Rosanne Tackaberry: 125

Michele Black/Great Backyard Bird Count: 64
David Boyle: 185
Mario E. Campos Sandoval/Costa Rica: 61
Kelly Colgan Azar: 166
Andrea Ferretti, from 'Sleeping Unsafely Tucked in to Conserve Energy in a Nocturnal Migratory Songbird,' *Current Biology*, 29, Issue 16, August 19, 2019, 2766–2772.
Jerry Friedman: 36B
Getty Images/Mike Bons: 33; Richard McManus: 77; Rasto Rejko: 53; Alexandre Shimoishi: 107
Jedidiah Gordon-Moran: 202
Bart Hardorff: 124
Peter W. Hills, worldbirdphotos.com: 171
Greg R. Homel/Natural Elements Productions: 183
Emily McKinnon & Kelsey Bell: 206
Dionne Miles: 127
Bret Nainoa Mossman: 13

Nature Picture Library/Karine Aigner: 170; Barrie Britton: 114; Robin Chittenden: 75; Lou Coetzer: 88; Russell Cooper: 84L; Tui De Roy: 119; Jasper Doest/Minden: 79T; Richard Du Toit/Minden: 29; Jack Dykinga: 5; Suzi Eszterhas: 84R; Angelo Gandolfi: 34; Steve Gettle/Minden: 167; Jen Guyton: 28; Erlend Haarberg: 189; Kerstin Hinze: 153; Tom Hugh-Jones: 201; Donald M. Jones/Minden: 138; Tim Laman/Nat Geo Image Collection: 99; Ole Jorgen Liodden: 45; Roy Mangersnes: 92, 110; Bence Mate: 2; Yva Momatiuk & John Eastcott: 210; Vincent Munier: 83; Greg Oakley/BIA/Minden: 67; Michel Poinsignon: 190; Roger Powell: 102; Marie Read: 65; Rosl Roessner/BIA/Minden: 150; Cyril Ruoso: 35; Andy Sands: 175; Gary K. Smith: 116; Matthew Studebaker/Minden: 55; David Tipling: 76, 80, 100; Markus Varesvuo: 22, 23; Visuals Unlimited: 37T; Gerrit Vyn: 215; Wild Wonders of Europe/López: 180T; Wild Wonders of Europe/Varesvuo: 70; David Williams/BIA/Minden: 178; Konrad Wothe: 179

Eric Nie: 213
Pius Notter: 20
Christina Riehl: 59, 62, 63
Dave Roach: 91
Laurie Ross/Tracks Birding and Photography: 49
Science Photo Library/Frans Lanting/Mint Images: 164

Shutterstock/Aaltair: 94; Agnieszka Bacal: 95; Bachkova Natalia: 73L; Neil Bowman: 17T; Steve Byland: 148; Tony Campbell: 27; Charlathan: 120; Clickmanis: 81; S Curtis: 169; Foto 4440: 130; Iliuta Goean: 180B; JMX Images: 154; Kajornyot Wildlife; Photography: 31; Cezary Korkosz: 87T; Brian Lasenby: 14; Sander Meertins Photography: 44; Leonie Ailsa Puckeridge: 136; Paul Reeves Photography: 17B; Tom Reichner: 187; Jean-Edouard Rozey: 38; Ryan S Rubino: 123; Super Prin: 79B

Claire Spottiswoode: 26L, 26R, 135, 165
Kenji Suetsugu: 30
Toshitaka Suzuki: 159
Keita Tanaka: 134
Wenfei Tong: 9L, 9R, 11, 56
Janske Van De Crommenacker: 142
Wikimedia Commons/J. J. Cadiz, Cajay: 87B; Zoologische Sammlung der Universität Rostock: 205
Ken Wilson: 39
Dr. Tzu-Ruei Yang, National Museum of Natural Science, Taiwan: 129
Wayne B. Young: 69
Christina Zdenek: 98

Every effort has been made to trace copyright holders and acknowledge the images. The publisher welcomes further information regarding any unintentional omissions.